Advances in Geopolymers

Edited by Pavlo Kryvenko
and Angel Palomo

Published in London, United Kingdom

Advances in Geopolymers
http://dx.doi.org/10.5772/intechopen.1004987
Edited by Pavlo Kryvenko and Angel Palomo

Contributors
Aline Figueirêdo da Nóbrega, Desalegn Abera, Dong Guo, Igor Rudenko, Jhennifer Larissa de Barros Silva, José Anselmo da Silva Neto, José Augusto Gomes Neto, Lan-Ping Qian, Marianne Saba, Oleksandr Konstantynovskyi, Pavlo Kryvenko, Wahib Arairo, Yun-Lin Liu, Zenamarkos Bantie

First published in London, United Kingdom, 2025 by IntechOpen
IntechOpen is the global imprint of INTECHOPEN LIMITED, registered in England and Wales, registration number: 11086078, 167-169 Great Portland Street, London, W1W 5PF, United Kingdom

For EU product safety concerns: IN TECH d.o.o., Prolaz Marije Krucifikse Kozulić 3, 51000 Rijeka, Croatia, info@intechopen.com or visit our website at intechopen.com.

British Library Cataloguing-in-Publication Data
A catalogue record for this book is available from the British Library

Advances in Geopolymers
Edited by Pavlo Kryvenko and Angel Palomo
p. cm.

This title is part of the Materials Science Book Series, Volume 4
Topic: Metals and Nonmetals
Series Editor: Chonghe Li
Topic Editor: Joan Josep Roa Rovira

Print ISBN 978-0-85014-969-2
Online ISBN 978-0-85014-968-5
eBook (PDF) ISBN 978-0-85014-970-8
ISSN 3049-8856

If disposing of this product, please recycle the paper responsibly.

IntechOpen Book Series

Materials Science

Volume 4

Aims and Scope of the Series

Materials science has always occupied an extremely high position in the human development process. As we explore the oceans of stars, various industries have put forward more stringent requirements for the performance of materials, forcing us to pay more and more attention to the development of new materials. At the same time, the formation of a data-driven scientific paradigm is dramatically shortening the development cycle of new materials. The huge data generated by synergistically combining theories, high-throughput experiments, high-throughput computation, and artificial intelligence is greatly contributing to our ability to utilize materials science to solve real-world problems. The three topics of this book series - Metals and Nonmetals; Composite Materials; and Surface Science - will address important areas of advancement in materials science. There will be a range of interesting works published under these topics.

Meet the Series Editor

Prof. Chonghe Li received his Ph.D. from the Chinese Academy of Sciences in 1995. From 1995 to 2000, he worked as a researcher at the Shanghai Institute of Metallurgy, Chinese Academy of Sciences, where he also served as director of the research laboratory. In 2000, he was appointed Professor at the Institute of High-Performance Computing in Singapore, where he worked on computation and simulation in materials science until 2004. Since then, he has been a professor at the School of Materials Science and Engineering, Shanghai University, China, as well as the director of the Shanghai Specialty Casting Engineering and Technology Research Center. Prof. Li's research focuses on titanium alloys, titanium-aluminum single crystals, intermetallic compounds, theoretical calculations, alloy design, and special refractory materials. His broad scientific expertise is well recognized by the scientific community around the world. He is a member of the editorial board of the journal Metals. As an author, he has published more than 200 peer-reviewed papers, 2 books, and over 40 patents.

Meet the Volume Editors

DSc Pavlo Kryvenko is currently a professor at the Department of Technology of Building Structures and Products and Director of the Scientific Research Institute for Binders and Materials of the Kyiv National University of Construction and Architecture. He received his Ph.D. from the Kyiv Civil Engineering Institute (1971) and DSc from the Kyiv Polytechnical Institute (1986). His research interests include the synthesis of alkali-activated cements and materials based on them. He has published over 20 books, over 1000 papers, and over 500 patents. He is a principal investigator in several national and international projects in Ukraine and abroad. He is a member of three RILEM Technical Committees on alkali-activated materials, a board member of three international journals, winner of the State Prize of Ukraine in Science and Engineering (2001), and supervisor of 6 DSc(Eng) and 23 Ph.D. theses.

Angel Palomo Sánchez holds a Ph.D. in Chemistry (1985) and has been part of the Spanish National Research Council (CSIC) research staff at the Eduardo Torroja Institute for Construction Science (IETcc) since 1981. He has extensive experience in construction materials research, focusing primarily on cement and concrete, including manufacturing, durability, and developing new binders. Angel Palomo Sánchez is the author or co-author of more than 170 scientific papers in listed journals and 30 in technical journals. He holds four patents for inventions associated with obtaining mineralized cement and alkaline binders. He has supervised six Ph.D. theses for professionals now engaging in scientific or industrial research. He has participated in 70 projects and research contracts, both as principal investigator and co-investigator. Additionally, he has co-supervised six Ph.D. theses. He is the author of 172 scientific articles published in SCI (Scopus)-listed journals, with 80% of them in Q1.

Contents

Preface

The increasingly growing population of the Earth and the need to save natural resources and energy and minimize carbon dioxide emissions over the past few decades have led to the development of new materials.

Geopolymers, a subclass of alkali-activated materials, are analogues of aluminosilicates that naturally occur due to chemical weathering and the interaction of the resulting products with the formation of zeolite-like minerals.

The processes of mineral formation occurring in nature were used to develop a class of inorganic polymer materials that possess a unique set of properties far exceeding those of organic polymers. Contrary to organic polymers, these inorganic polymers are environmentally friendly and help protect human health, particularly in case of fire (they do not produce toxic emissions). Above all, activating alkalis provides the opportunity to involve many industrial wastes and by-products in the production process, helping address environmental problems.

The defining characteristic of geopolymer is the binding phases comprised of an alkali aluminosilicate gel, with aluminium and silicone linked in a three-dimensional tetrahedral gel framework that is relatively resistant to water. In other words, the primary method for synthesizing geopolymers involves combining an alkaline solution with reactive aluminosilicate powder, particularly metakaolin (as discussed in the chapters "Optimization of Alkali-Activated Concrete Dosage Based on Metakaolin from the Brazilian Northeast" and "Geopolymers in Civil and Environmental Engineering Applications"), or fly ash.

This results in the formation of a disordered alkali aluminosilicate gel phase, known as the geopolymeric gel-binder phase of system $Na_2O(K_2O)$-Al_2O_3-SiO_2-H_2O.

By changing the ratios of the listed oxides and using modifiers containing Ca^{2+} or Mg^{2+} ions (Chapter "Optimization of Alkali-Activated Concrete Dosage Based on Metakaolin from the Brazilian Northeast"), it is possible to synthesize the specified properties of such binders for various applications, such as dispersed reinforced concrete (Chapter "Performance Enhancement of Geopolymers through Natural Fiber Reinforcement: Sustainable Alternative Construction Materials"), thermal insulating materials (Chapter "Foaming Processes and Properties of Geopolymer Foam Concrete"), as well as in civil and environmental applications (Chapter "Geopolymers in Civil and Environmental Engineering Applications").

This book covers recent activities and the accumulated knowledge of researchers from multiple disciplines worldwide, providing a comprehensive overview of advancements in the synthesis and characterization of geopolymers. It includes a broader range of raw materials that can be used and explores their applications in various fields.

Summarizing general research practices and principles, including case studies drawn from the firsthand experiences of scientists, engineers, regulators, and educators, this book can be used in advanced environmental engineering courses. It also serves as a valuable reference for emerging engineers, technology vendors and buyers, laboratory technicians, scientists, environmentalists, policymakers, and managers.

Pavlo Kryvenko
Scientific Research Institute for Binders and Materials,
Kyiv National University of Construction and Architecture,
Kyiv, Ukraine

Angel Palomo
Eduardo Torroja Institute for Construction Sciences,
Madrid, Spain

Chapter 1

Optimization of Alkali-Activated Concrete Dosage Based on Metakaolin from the Brazilian Northeast

José Anselmo da Silva Neto, Jhennifer Larissa de Barros Silva, José Augusto Gomes Neto and Aline Figueirêdo da Nóbrega

Abstract

Alkali-activated structural concretes that are both pumpable and machinable were synthesized utilizing a low-reactivity metakaolin characterized by elevated levels of quartz and iron. The investigation into the effects and interactions of alkaline concentration (NaOH—8, 10, 12, 14, and 16 M) and the ratio of alkaline reagents (Na_2SiO_3/NaOH—1.0, 1.5, 2.0, 2.5, and 3.0) on the material properties was conducted employing a 2^k full factorial design. An increase in alkaline concentration (NaOH), when coupled with a decrease in the ratio of alkaline reagents (Na_2SiO_3/NaOH), expedited both the initial and final setting times of the mixtures. Elevated alkaline solution concentrations led to a diminished slump, swifter hardening, and enhanced mechanical strength; however, surpassing the threshold of 14 M for the same Na_2SiO_3/NaOH ratio resulted in a 12.4% decline in compressive strength at 28 days. The factorial design facilitated the identification of the optimal dosage for the material's production, revealing that an alkaline concentration (NaOH) of 10 M and a Na_2SiO_3/NaOH ratio of 2.5 are requisite. The evaluation of various potential combinations enabled the attainment of a material possessing properties that are conducive to structural applications.

Keywords: low reactivity, quartz, iron, factorial planning, dosage

1. Introduction

Alkali-activated concretes formulated from metakaolin have emerged as a subject of considerable scholarly investigation as a viable alternative to Portland cement concrete, aimed at mitigating energy consumption and global carbon dioxide (CO_2) emissions [1–3]. Nevertheless, critical considerations regarding the implementation of geopolymer concretes in industrial process facilities and field operations encompass their rapid setting characteristics and insufficient consistency for optimal release and effective densification, which directly impacts the void volume and porosity of the geopolymer; thus, it also affects its ultimate compressive strength [4].

Metakaolins characterized by significant quartz and iron impurities, coupled with a diminished surface area, typically exhibit low reactivity and, as a consequence, are less desirable in commercial applications within the realm of geopolymers [5]. Without further elaboration, it is posited in Silva et al., Brazilian Association of Technical Standards and Lemougna et al. [5–7] that metakaolin exhibiting a quartz content of less than 50%, a minimal iron proportion generally not exceeding 3%, and an elevated surface area surpassing 15 m^2/Kg is inclined to demonstrate heightened reactivity, rendering it more desirable for use in geopolymers.

In this context, the authors opted to utilize locally sourced metakaolin (originating from the northeastern region of Brazil), thereby mitigating the extensive transportation of materials via roadways within Brazil. This particular metakaolin exhibits three noteworthy chemical properties from the perspective of geopolymerization kinetics, aimed at prolonging the setting durations and minimizing the viscosification of the geopolymer matrix: the diminutive specific surface area, the presence of crystalline quartz, and the concentration of iron [6]. This metakaolin, characterized by elevated levels of iron and quartz along with a reduced specific surface area, has been designated as MKFeQ in this study.

Assessing the effects of alkaline concentration (NaOH) [8] and the proportions of alkaline reagents ($Na_2SiO_3/NaOH$) within this framework is imperative [9–11], not solely owing to the varying properties of the precursor utilized to regulate the setting duration but also to enhance comprehension of the system, which is particularly critical given the absence of an established standard dosing protocol, as is prevalent in Portland cement concrete (ABCP, IPT, and others) [12, 13].

Recent studies have highlighted the significant role of Fe_2O_3 in the structural formation and mechanical properties of alkali-activated cements, particularly those based on metakaolin. The presence of iron oxide influences the geopolymerization process by enhancing the formation of iron-silicate binding gels, which contribute to the overall strength and durability of the material. According to Silva Neto et al. [14], the amalgamation of iron and quartz within the constituent materials has the potential to extend the duration of the polymerization reaction, thereby enhancing the mechanical integrity of the geopolymer concrete as time progresses. The iron-silicate binding gel generated throughout the chemical interaction is essential for the mechanical robustness of the material. Furthermore, Ngnintendem et al. [15] observed that the presence of iron minerals in metakaolin-based geopolymers can lead to enhanced compressive strength and microstructural properties. Similarly, Gomes et al. [16] demonstrated that iron's distribution within geopolymers significantly affects the final properties of the material, making it essential to control Fe_2O_3 content during the formulation to optimize performance.

The role of temperature in the development of strength and the evolution of the structure of alkali-activated concretes is crucial. During the geopolymerization process, temperature directly influences the reaction rate and the formation of the three-dimensional matrix that characterizes these materials. Higher temperatures increase the solubility of aluminosilicates, accelerating the polymerization of monomers into a denser and stronger structure. However, excessively high temperatures can damage the matrix integrity, leading to the formation of undesirable phases such as mullite, which compromises mechanical strength and increases material porosity.

Furthermore, the curing regime under different temperature conditions significantly affects the microstructural and mechanical properties of the concrete. Curing at moderately high temperatures (around 60°C) has proven effective in obtaining geopolymers with excellent strength and low porosity by promoting more complete geopolymerization and forming a homogeneous structure. Conversely, curing at very high temperatures or for extended periods can cause excessive dehydration and the formation of microcracks, which deteriorate the material's properties in the long term. Therefore, controlling temperature during the curing process is essential to optimizing the final characteristics of alkali-activated concretes, ensuring a balance between strength, durability, and dimensional stability.

Regarding the concentration of alkali (NaOH), while it has been documented that alkali-activated composites derived from metakaolin have employed concentrations ranging from 8 to 16 M NaOH [17, 18], an escalation in the alkaline concentration to 12 M has been shown to enhance the mechanical properties of alkali-activated materials based on metakaolin [3, 19, 20].

The role of Fe_2O_3 in the development of mechanical strength and structural formation of alkali-activated concretes is significant, particularly when considering varying curing temperatures. Fe_2O_3 contributes to the formation of iron-silicate gels, which enhance the material's compressive strength by promoting denser and more interconnected aluminosilicate networks. The curing process at different temperatures plays a critical role in optimizing these properties. For instance, curing at moderate temperatures, such as 60°C, has been shown to accelerate the geopolymerization process, resulting in a more uniform and robust microstructure. However, excessively high curing temperatures can lead to the formation of undesirable phases, such as mullite, which may compromise the material's mechanical performance. Therefore, the careful control of Fe_2O_3 content and curing conditions is essential for achieving optimal performance in alkali-activated concretes.

Elevated concentrations of sodium hydroxide (NaOH) can expedite polymerization reactions, leading to saturation and constraining the interaction among polymerized molecules [21–23], which consequently results in an augmentation of the solution's viscosity [22], a diminishment in setting durations [20], and the retention of unreacted precursor substances within the final product [22], accompanied by a decline in mechanical properties and an increase in porosity levels [3, 24, 25], as well as an elevated likelihood of efflorescence occurrence [26, 27]. While modifications in mixing ratios and materials have postponed this configuration, they have concurrently induced a decrease in mechanical properties [8, 28].

Conversely, the augmentation of alkaline reagents (Na_2SiO_3/NaOH) has been documented to enhance the polymerization process [29, 30]. The predominant ratios of Na_2SiO_3/NaOH ranging from 1.0 to 3.0 have been identified in composites activated by alkali and based on metakaolin [17, 18, 31]. Indeed, Na_2SiO_3/NaOH ratios approaching 3.0 are frequently reported to be advantageous for the attainment of superior mechanical strength [19]. Nevertheless, it is imperative to exercise caution regarding the diminished workability that accompanies an increase in the Na_2SiO_3/NaOH ratio [28]. Decreases of as much as 30% in slump are noted in the frustum due to the escalation of the Na_2SiO_3/NaOH ratio from 1.5 to 2.5 in alkali-activated concretes derived from metakaolin [32–34], thereby complicating workability because of the viscous nature of the mixture, which subsequently leads to a reduction in strength [35, 36].

Therefore, the objective of this investigation is to enhance the understanding of alkali-activated concretes that utilize metakaolin as a primary constituent. The enhancement of scholarly insights pertaining to metakaolin characterized by elevated levels of quartz and iron impurities, along with a reduced specific surface area (MKFeQ), holds significant relevance in the current discourse. In light of the aforementioned context, this study addresses a critical deficiency in this domain of knowledge by providing empirical findings.

2. Materials and methods

2.1 Materials

For the production of concrete, aggregates, water, sodium silicate, sodium hydroxide, and MKFeQ were used, the latter being used as the primary source of silica and alumina. The activating solution consisted of a sodium silicate solution (Na_2SiO_3), used as an extra source of silica, and sodium hydroxide (NaOH), the alkaline medium necessary for geopolymerization, accompanied by water. The properties provided by the manufacturer for silicate and sodium hydroxide are shown in **Tables 1** and **2**, respectively. To synthesize concrete, coarse and fine aggregate were added to the alkali-activated paste.

The materials were characterized in terms of physical, chemical, and mineralogical aspects. In the physical analysis, the particle size and specific mass of the aggregates and fines were verified. In the chemical analysis, the X-ray fluorescence test (XRF) was carried out, and in the mineralogical analysis, by X-ray diffractometry (XRD).

Figures 1 and **2** show the particle size curves of the fine and coarse aggregate, respectively. The fine aggregate had a maximum diameter of 2.40 mm and a fineness modulus of 2.47, being classified by NBR NM 248 [37] as medium sand. The granulometric curve of the sand used behaved practically between the lower and upper limits of the optimum zone. The coarse aggregate had a maximum diameter of 25.00 mm and was delimited between the lower and upper usable zones.

The particle size distribution of metakaolin was determined utilizing the laser beam diffraction technique, employing a Cilas 1180 laser granulometer, to assess

Determination	Specification	Result
Physical state (25°C)	Liquid	Liquid
Coloring	Colorless	Colorless
Alkalinity: Na_2O (%)	16.00–19.00	16.13
Silica: SiO_2 (%)	33.00–38.00	34.20
Total solids (%)	50.00–57.00	50.33
Water: H_2O (%)	—	49.67
Density	1.63–1.67	1.63

Source: Pernambuco Química S/A (2023).

Table 1.
Characteristics of sodium silicate.

Characteristics	Inferior limit (%)	Upper limit (%)	Result (%)
Sodium Oxide: Na_2O	—	0.040	0.040
Iron Oxide: Fe_2O_3	—	0.006	0.006
Purity	98.00	—	98.50

Source: Bandeirante Brazmo (2023).

Table 2.
Characteristics of sodium hydroxide.

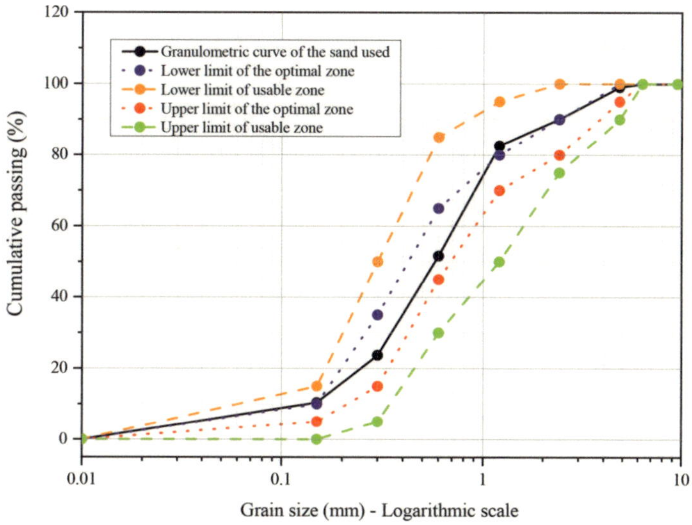

Figure 1.
Fine aggregate granulometric curve.

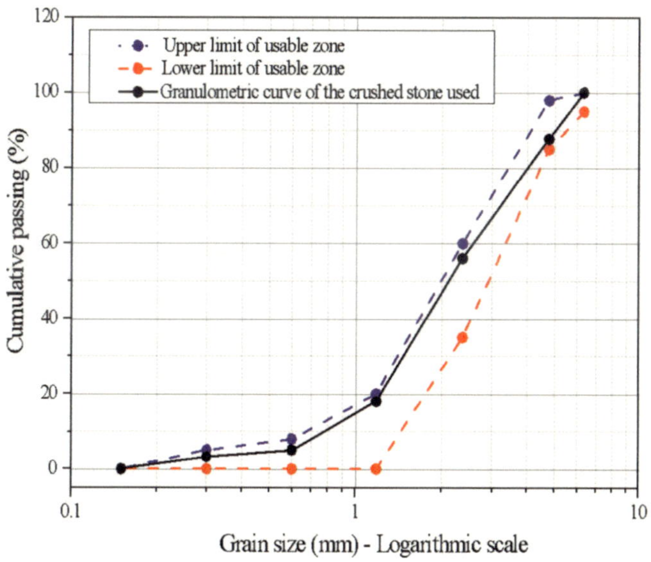

Figure 2.
Coarse aggregate granulometric curve.

the distribution of the material across a size spectrum ranging from 0.5 to 500 μm. According to the particle size analysis of metakaolin, 10% of the solid constituents are smaller than 0.99 μm; 50% are below 3.80 μm, and 90% are less than 32.13 μm. The mean diameter of the particles was recorded at a value of 10.41 μm. It is evident that the predominant concentration of fine metakaolin particles is located within the range of 0.48–32.13 μm (**Figure 3**). As stated in Zhang et al. [38], this average particle diameter of metakaolin is pivotal in facilitating the filler effect and in the formation of nucleation sites within alkali-activated matrices.

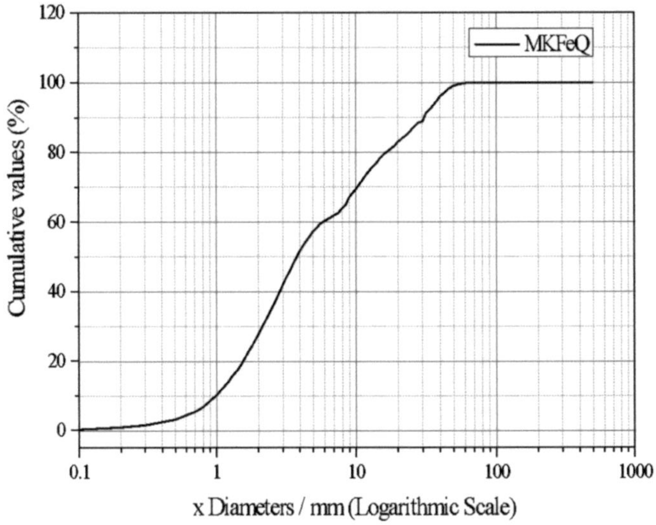

Figure 3.
Metakaolin particle size curve.

The specific gravity of the aggregates was determined using the Archimedes principle, following NBR NM 52 [39] for fine aggregate and NBR NM 53 [40] for coarse aggregate. The fine and coarse aggregates had a specific mass of 2.67 and 2.79 g/cm^3, respectively. The specific mass of metakaolin complied with NBR 16605 [41], establishing two readings, and the average of these readings was the value considered, which was 2.58 g/cm^3.

Regarding the diagnosis of the chemical composition of metakaolin (FRX), the sample was sieved through mesh n°. 200 mesh, then tablets were compressed at a load of 80 kN and placed in the sample holder of the Shimadzu EDX-720 equipment. According to the result obtained in **Table 3** and in accordance with NBR 12653 [42], MKFeQ met the requirements as Class N pozzolanic material, as it presented a SO_3 content lower than 4% and the sum of SiO_2, Al_2O_3, and F_2O_3 greater than 70%.

The role played by Fe_2O_3 (iron oxide) in the formation of the structure of geopolymers is fundamental in determining the integrity and final properties of the material. The presence of Fe_2O_3, especially in high concentrations, directly influences the polymerization reaction, contributing to the stability of the three-dimensional structure formed during the geopolymerization process. In metakaolin-based geopolymers, Fe_2O_3 can act as a structure modifier, promoting the formation of stronger bonds between silicon and aluminum atoms, resulting in a denser matrix with improved mechanical properties. However, the amount of Fe_2O_3 must be carefully

Material	SiO_2 (%)	Al_2O_3 (%)	Fe_2O_3 (%)	CaO (%)	MgO (%)	SO_3 (%)	Na_2O (%)	K_2O (%)	TiO_2 (%)	LOI* (%)
MKFeQ	53.65	31.79	9.90	0.12	1.17	0.08	0.20	0.65	1.84	0.60

**LOI: Mass Loss.*

Table 3.
Chemical composition of MKFeQ.

controlled, as excessive concentrations can lead to the formation of undesirable secondary phases that may compromise the durability and strength of the material.

Additionally, Fe_2O_3 contributes to the distinct coloration in the geopolymer matrix, which can indicate the distribution of iron within the structure. Its presence can also influence the reactivity of metakaolin during alkaline activation, altering the dissolution rate of aluminosilicates and, consequently, the curing speed and the evolution of strength over time. The interaction between Fe_2O_3 and other components, such as Na_2O, should be considered when formulating geopolymers, as it affects ion solubility and the final polymeric network formation.

For the mineralogical analysis by X-ray diffraction, the sample was sieved through mesh N°. 200 mesh and subsequently deposited in the sample holder, undergoing slight compression before being deposited in the Siemens Bruker model D5000 equipment using CuK α wavelength radiation $\lambda = 1.5418$ with X-rays at 30 kv and 30 mA, reading speed of 1°/min in a range of 5–70° 2θ at an angular step of 0.02° 2θ. To identify the peaks, the X'Pert software was used. HighScore Plus 2.0. **Figure 4** shows the MKFeQ diffractogram.

According to **Figure 4**, it is possible to observe some intense peaks of MKFeQ, mainly composed of crystalline phases of feldspar, muscovite, and quartz. When these minerals appear in kaolinite clay, they can be harmful to its quality, affecting some important properties such as whiteness, viscosity, and abrasiveness [43]. Another important factor to be analyzed in **Figure 4** is the amorphous halo present between 22° and 27° 2θ, although it is not of great intensity, but its presence is common in pozzolans and, generally, indicates a certain reactive potential of pozzolanic materials.

2.2 Methods

After characterizing the materials, it was possible to carry out the 2^k factorial experimental design (k = 2) through the complete linear models using the Statistica 10.0 software. The values used were based on research by Zhang et al. [44]. Factorial

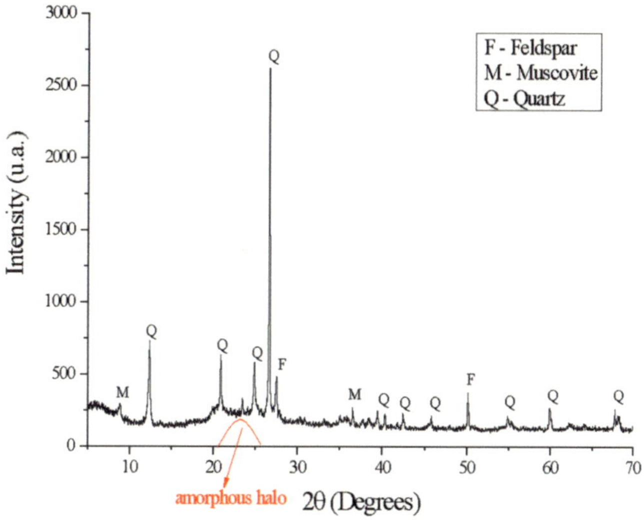

Figure 4.
MKFeQ X-ray diffractogram.

Variable	-1	0	+1
NaOH (M)	10	12	14
Na_2SiO_3/ NaOH	1.5	2.0	2.5

Table 4.
Variable levels.

planning using linear models includes defining the study levels of the variables (−1, 0, +1), shown in **Table 4**.

Synchronously, the setting time, slump, and resistance to axial compression of these composites were evaluated in relation to the extreme values with regard to the central points of each variable obtained from Zhang et al. [44], following the same methodology as replicates, in order to obtain the average of the response variables, as shown in **Table 5**.

The composition of the chosen mass mix was 1:2:3 (metakaolin: sand: gravel) based on previous work by Melo [45] and because it results in good fresh and hardened characteristics. The solid/liquid ratio was 0.80; this ideal value was also obtained by Perez-Cortes and Escalante-Gracia [46] in his research. The production of the test specimens was carried out in accordance with NBR 5738 [47]. However, except for the curing configuration, where thermal curing took place at a temperature of 60°C for a period of 48 hours.

After the thermal curing process at 60°C for 48 hours, the hardened specimens were subjected to a series of tests to evaluate their properties. The axial compressive strength was measured according to NBR 5739 [48], using cylindrical specimens of 100 mm x 200 mm. Additionally, microstructural analyses were conducted using scanning electron microscopy (SEM) to assess phase distribution and material density. Furthermore, Fourier Transform Infrared Spectroscopy (FTIR) was performed to identify changes in chemical bonds present in the samples after alkaline activation. These tests provided a comprehensive understanding of the physical and chemical properties of the produced geopolymers.

Alkaline activated pastes were evaluated according to Brazilian Association of Technical Standards [49] and the consistency of the concretes through the cone slump test according to NBR NM 67 [50]. On the other hand, regarding the hardened property, the resistance to axial compression was verified by NBR 5739 [48]. Fifteen cylindrical specimens (100 mm x 200 mm) were made for each type of mixture produced, obtaining the arithmetic mean of three specimens after 28 days.

In statistical modeling, the response surfaces were established in the diagram by two vertices (independent variables), modifying the last vertex with the response

Extreme points			
TRAITS	Codification	NaOH (M)	Na_2SiO_3/ NaOH
	T-8-2.0	8	2.0
	T − 16-2.0	16	2.0
	T − 12 − 1.0	12	1.0
	T-12-3.0	12	3.0

Table 5.
Combination of coded parameters of extreme points.

variable (dependent variable) defined with p-values equal to 0.05. Finally, the models were validated using a diagram with the predictive and observed values to compare them, as well as through a calculation based on an analysis of variance table (ANOVA), analyzing the F-Test and R^2.

The specimens utilized for the scanning electron microscopy examination (FEI-Quanta 450, 20 kV, 250x) were fabricated with dimensions measuring (3 x 3 x 1) cm and were extracted from the central region of the samples employing a low-speed saw equipped with a diamond disk lubricated by mineral oil. Subsequent to the cutting procedure, the specimens were subjected to immersion in isopropyl alcohol under ultrasonic treatment for a duration of 24 hours, followed by drying in an oven at a temperature of 35°C for an additional 24 hours, after which they were encapsulated in resin and meticulously polished using progressively finer grades of sandpaper: 400, 600, 1000, 1200, 1500, and 2000 grit, in sequence. Upon completion of the polishing process, the samples were coated with a conductive layer of carbon and were stored in a desiccator at ambient temperature until such time as analysis was conducted.

The Fourier Transform Infrared Spectroscopy (FTIR) analysis was conducted with the objective of elucidating alterations in the chemical bonds inherent in the precursor (MKFeQ) and subsequent to the exposure of the activating solutions to the alkali-activated concretes following a duration of 28 days. The specimens were subjected to fragmentation and sieving (# 200 μm), resulting in the acquisition of a fine particulate material. The examination was performed utilizing a Shimadzu spectrophotometer (IRPrestige-21) for transmittance measurements at a velocity of 0.2 cm/s and a resolution of 4 cm^{-1} across a spectral range spanning from 4000 to 400 cm^{-1}. For the execution of the analysis, potassium bromide (KBr) was employed in conjunction with the samples, maintaining identical mass proportions for each assessment in the presence of ambient atmospheric conditions.

3. Results

3.1 Fresh property

Figure 5 shows the results of the test regarding the final and initial setting times mutually.

From the image above, it is observed that the increase in alkaline concentration (NaOH), accompanied by the reduction in the ratio of alkaline reagents (Na_2SiO_3/NaOH), accelerated the dissolution rate of the source materials. This accelerated kinetics is perceived in the T-10-2.5 mix compared to the T-16-2.0, reducing the initial and final setting times by 100 and 115 minutes, respectively.

Furthermore, it can also be seen that the reduction of the Na_2SiO_3/NaOH ratio from 2.5 to 1.5, emphasizing the traces T-10-2.5 and T-10-1.5 for the same concentration alkaline (10 M), reduced the initial and final setting time by 10 and 5 minutes mutually. On the other hand, when further lowering the Na_2SiO_3/NaOH ratio from 3.0 to 1.0 for a higher alkaline concentration (12 M), the trace T-12-3.0 compared to T-12-1.0 showed a reduction in initial and final setting times of 35 and 40 minutes, respectively. According to Davidovits [3], the attenuation of the SiO_2/Na_2O ratio present in the activating solution accelerates the reaction rate and the hardening of aluminosilicates.

The ratio of Na_2SiO_3 to NaOH, quantified at 2.0 and denoted in traces T-8-2.0 and T-16-2.0, resulted in a reduction of both the initial and final setting times of the

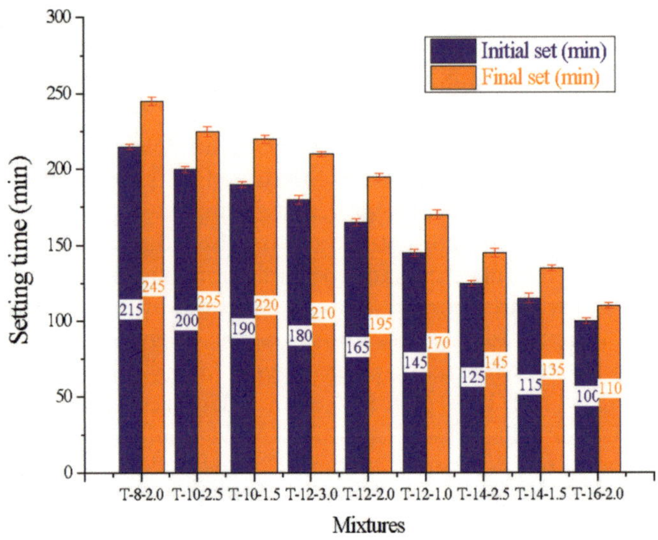

Figure 5.
Setting times of MKFeQ-based alkali-activated pastes.

formulations, which decreased to 115 and 135 minutes, respectively. Conversely, a decrease in alkalinity from 16 to 12 M (T-12-2.0) led to an increase in the initial and final setting times by 65 and 85 minutes, respectively. This observed phenomenon of reduced setting time can be attributed to the enhanced dissolution of Si^{4+} and Al^{3+} ions from the precursor materials at elevated pH levels, as well as the alteration in the silica modulus (SiO_2/Na_2O) of the dual activators comprising both hydroxide and silicate [48–50].

NBR 7212 [51], for the execution of batched concrete in a plant, stipulates the maximum time for it to be completely unloaded (applied) on site at 150 minutes. Therefore, only traits T-14-1.5, T-14-2.5, and T-16-2.0 do not qualify as viable for such an application.

After preparing the traces for molding the test specimens, the slump analysis of the mixtures was immediately carried out using the slump test. The abatement results can be seen in **Figure 6**.

In **Figure 6**, it is noticeable in the mixtures that when using the same alkaline concentration of 12 M and varying the $Na_2SiO_3/NaOH$ ratio (1.0–3.0), there was a reduction in spreading of 72 mm, respectively, in the mixtures. T-12-1.0, T-12-2.0, and T-12-3.0. In the work Xie et al. [52], the same behavior is reported: when increasing the ratio of alkaline reagents from 1.0 to 3.0 in the mixture, the slump was reduced by 82 mm for an alkaline concentration analogous to 10 M. This happens because the solution of sodium silicate as an auxiliary activator is more viscous than sodium hydroxide [53].

Conversely, concerning the augmentation of alkaline concentration at a consistent $Na_2SiO_3/NaOH$ ratio, a diminished reduction was also observed, as illustrated by the trace T-8-2.0 in comparison to T-16-2.0, which exhibited a decline from 107 to 87 mm. This phenomenon can be ascribed to the enhanced viscosity of sodium silicate concomitant with the escalation of sodium hydroxide, as well as to the accelerated leaching of aluminosilicates from the parent material in an elevated pH environment [54].

In general, pumpable concrete must have good workability, that is, the slump must be greater than 70 mm, with values between 80 and 100 mm being the most

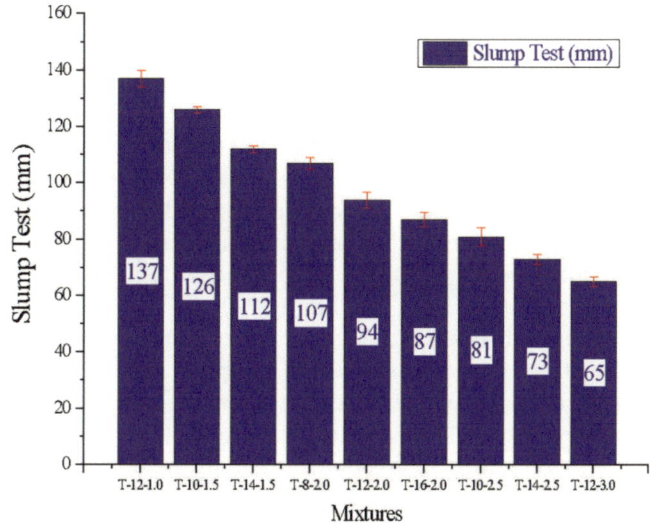

Figure 6.
Slump test of alkali-activated composites.

recommended. In view of this, the only traits that are not suitable for this application would be T-12-3.0 and T-14-2.5.

3.2 Hardened property

Figure 7 shows the axial compression resistance at 28 days of alkali-activated concrete.

From the traces T-8-2.0 and T-12-2.0 for the same Na_2SiO_3/NaOH ratio, the beneficial effect of increasing the alkaline concentration at 28 days can be seen, that is,

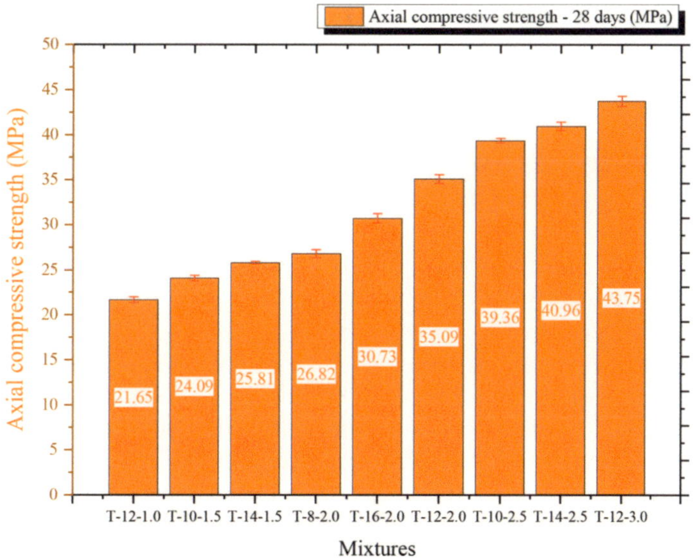

Figure 7.
Resistance I'm compression up to 28 days.

indicating an increase in compressive strength of approximately 8.3 MPa. This same behavior is reported by Chithambaram et al. [55], because due to the high content of cyclic silicate and because it is a lower alkaline concentration, it makes the condensation process difficult, resulting in lower resistances.

In theory, this increase proved to be beneficial in the study, but it was observed that when exceeding the limit of 14 M for the same ratio $Na_2SiO_3/NaOH$ = 2.0, such as at 28 days, the T-16-trace 2.0, there was a reduction of 12.4% for compressive strength and 11.5% for tensile strength when compared to the T-12-2.0 mix. This excessive increase in alkaline concentration reduced the stability of the oligomeric silicate particles (such as $Si_4O_8\ (OH)_6^{-2}$ and $Si_4\ O_8\ (OH)_4^{-4}$) and the equilibrium of these particles made the shift toward the formation of mononuclear silicate (such as $SiO_2\ (OH)_2$ and $SiO\ (OH)^{-3}$), as a consequence, the generation of excess $(OH)^-$. These excess hydroxide ions, in turn, reduce the effectiveness of the polycondensation process and, subsequently, reduce mechanical resistance [56].

From an alternative viewpoint, the enhancement in the $Na_2SiO_3/NaOH$ ratio from 1.0 to 3.0 at an alkaline concentration of 12 M proved to be conducive to the augmentation of mechanical strength, attributable to the degrees of silica and alumina dissociations of MKFeQ, in addition to the soluble silica content present in the sodium silicate solution when interacting with the sodium hydroxide solution within the specified molarity range. For instance, the samples T-12-1.0 and T-12-3.0, which exhibit a molarity corresponding to 12 M and where the ratio of $Na_2SiO_3/NaOH$ is varied from 1.0 to 3.0, demonstrated a compressive strength exceeding 22.1 MPa at the 28-day mark.

Given this scenario, and in accordance with NBR 8953 [57], all concrete produced was classified as having structural applicability, as it was between classes C20 and C40 at 28 days.

It is also pertinent to note that the concurrent presence of iron and quartz within the raw materials can extend the duration of the geopolymerization reaction when subjected to alkaline solutions, thereby enhancing the mechanical properties of the concrete. This phenomenon can be attributed to iron's considerable atomic mass and its comparatively larger atomic radius in relation to silica and alumina, which renders it more reactive in the context of the activating solution. Furthermore, quartz exhibits a minimal surface area, which implies a reduced interface available for interaction with the activator solution. The coexistence of iron oxide and quartz has been linked to an extended curing period, which in turn contributes to an elevation in the compressive strength of the geopolymer concrete. The iron-silicate binding gel generated during the chemical reaction is fundamental to the mechanical integrity of the material [5, 7].

3.3 Microstructural analysis

In **Figure 8**, it is observed that increasing the alkaline concentration for the same alkaline reagent ratio of 2.0 is beneficial in the mechanical resistance of the alkali-activated metakaolin matrices up to a certain limit, thus confirming the reason for the hardened property results. According to Kamseu et al. [8], alkali is characterized by promoting the solubility and dissolution of aluminosilicates, and normally the increase in alkaline concentration results in composites with superior mechanical resistance.

According to the data presented in **Figure 8**, for the mixture exhibiting an alkaline concentration of 8 M, there was a lack of complete dissolution of the aluminosilicates;

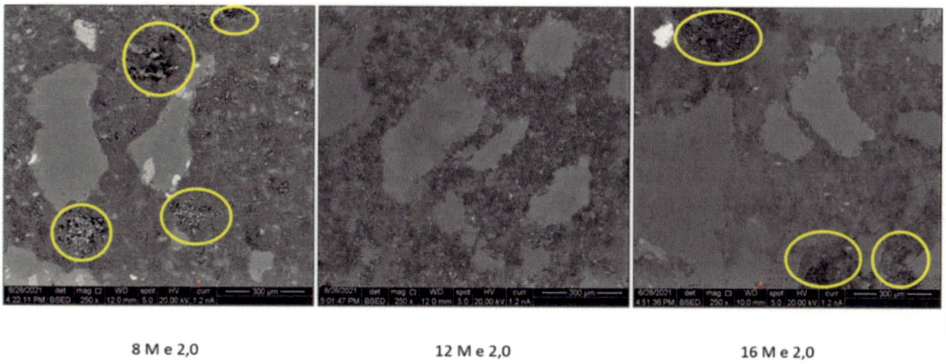

Figure 8.
Increase in alkaline concentration for the same ratio of alkaline reagents of 2.0 (a) 8 M (b) 12 M and (c) 16 M.

however, the elevation of the concentration from 8 to 12 M illustrates a more compact structural formation, with a portion of the resultant product demonstrating enhanced reactivity. Nonetheless, upon further augmentation of this concentration to 16 M, the dissolution appeared to be illusorily reinstated in an incomplete manner, akin to the sample at 8 M, albeit characterized by a smoother and more compact structure, which indicates a favorable response resulting from this increase in hardened properties.

Figure 9 illustrates the efficacy of augmenting the ratio of alkaline reagents while maintaining a constant alkaline concentration of 14 M. In the majority of instances, the escalation in the proportion of alkaline reagents is congruent with the mechanical strength of alkali-activated materials, as it promotes the polymerization process [38].

The increase in the ratio of alkaline reagents from 1.5 to 2.5 for the same alkaline concentration of 14 M (**Figure 9**) resulted in a more compact and dense structure, indicating a benefit in the polymerization process with the increase in this variable in response.

Figure 10 presents the infrared spectra of activated alkali concretes produced using MKFeQ up to 28 days.

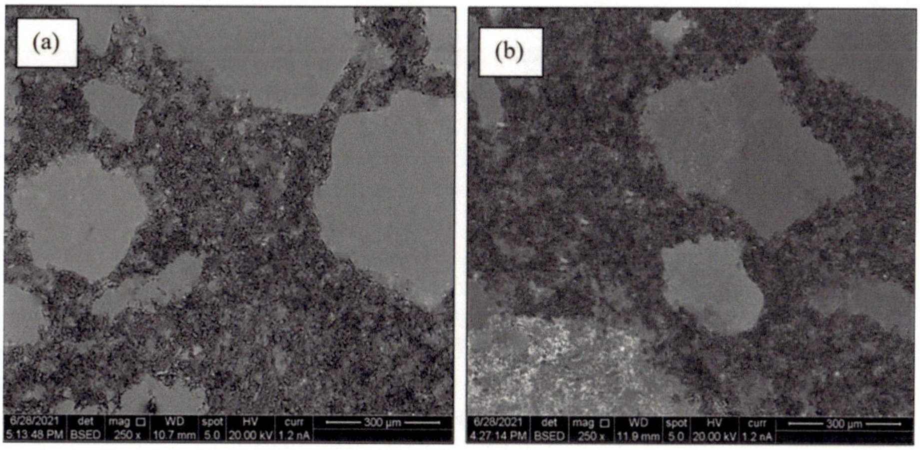

Figure 9.
Different ratios of alkaline reagents (Na_2SiO_3/NaOH) (a) 1.5 (b) 2.5.

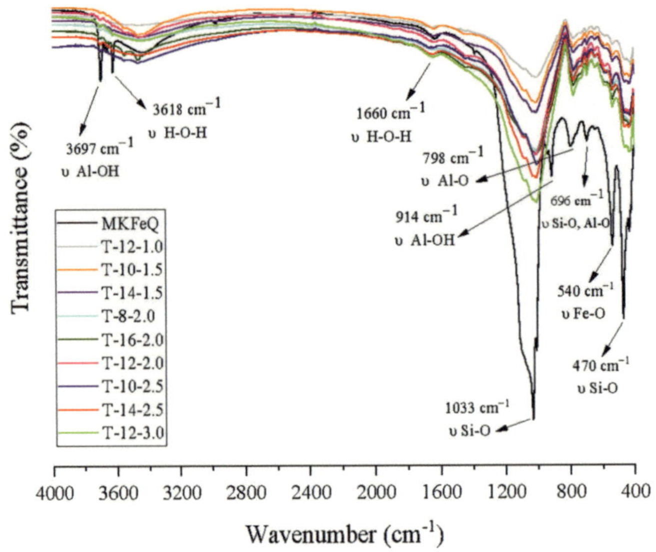

Figure 10.
FTIR of alkali-activated concrete produced using MKFeQ.

According to the results obtained, it is verified that the material presents a more accentuated presence of bands in the wavelength range from 1700 to 400 cm^{-1}, which can be attributed to the intrinsic characteristics of the material. It can be highlighted the presence of ban-related I'm vibrations of the OH group at 3697 cm^{-1}, referring to I'm Al-OH bonds and asymmetric stretching of the group present in water emissions, poorly adsorbed on the surface of the material, and bands in the region of 1660 and 798 cm^{-1}, attributed to I'm Hello group vibration and Si-O group length, respectively.

It is crucial to emphasize that the spectral band observed at 1033 cm^{-1}, which is ascribed to Si-O-Si groups within MKFeQ, exhibited a notable shift toward wavenumbers approximately near 1000 cm^{-1}, as evidenced in the spectra of geopolymerized specimens. This observed displacement can be attributed to the chemical interactions occurring during the geopolymerization process, specifically due to the substitution of silicon atoms with aluminum, resulting in the formation of Al-O-Si structures, thereby supporting the findings presented in previous investigations conducted by Brazilian Association of Technical Standards and Lizcano et al. [6, 58].

The diminution of the iron band to a value of 540 cm^{-1} may suggest that the iron contained within the initial sample underwent a chemical transformation during the synthesis of the geopolymer matrices. This phenomenon may arise due to the presence of alkaline ions within the solution, which are capable of solubilizing iron and facilitating its interaction with other constituents of the sample. The reduction of the Si-O and Al-O bands to a level of 696 cm^{-1} may imply that the concentration of these compounds within the sample has also diminished subsequent to exposure to the alkaline solution [13].

It is also feasible to underscore the pre-existing spectral bands within the range of 600–800 cm^{-1}, which are associated with the presence of Si-O-Al, thereby signifying the existence of crystalline phases, as corroborated by Brazilian Association of Technical Standards, Zhang et al., and Fang et al. [6, 38, 53]. The manifestation of these spectral bands can be elucidated through the examination of the precursor X-ray diffraction (XRD) results (MK), employed in this study, which reveals the presence of crystalline phases.

3.4 Statistical modeling

Table 6 presents the non-linear regression models that incorporate statistically significant parameters at a 95% confidence level ($p < 0.05$), along with the calculated F ratio (F-Test) and the coefficient of determination (R^2).

The coefficients highlighted in bold within the regression models presented in **Table 5** (responses) demonstrate statistical significance at the 95% confidence level. Consequently, based on the equations delineated in **Table 5**, it is evident that the first-order models exhibit statistical significance across all evaluation responses, as they possess an F-Test value exceeding 1.

The coefficient of determination, denoted as R^2, oscillates between the values of 0 and 1 and is occasionally represented in percentage format. In this context, it quantifies the proportion of variability in the data that is elucidated by the linear regression model. Consequently, an increase in the R^2 value signifies a greater explanatory power of the linear model, thereby indicating a superior fit to the dataset under consideration.

Statistically significant models were employed to construct response surfaces aimed at optimizing both the fresh and hardened properties of concrete. Consequently, to streamline the responses, only the principal characteristics of the concrete were assessed, specifically the final setting time, slump, and compressive strength at 28 days, as depicted in **Figure 11 (a)** and **(b)** and **Figure 12**, respectively.

Figure 11 (a) clearly shows the inversely proportional influence of the alkaline concentration (NaOH) and the directly proportional influence of the $Na_2SiO_3/NaOH$ ratio. It is also inferred that when fixing the independent variable (NaOH) in the response at level − 1 (10 M), it can be seen that for the value of $Na_2SiO_3/NaOH$ at level + 1 (2.5), one obtains the maximum final setting time value of 225 minutes.

Simultaneously, it is observable in **Figure 11 (b)** that the influence of the $Na_2SiO_3/NaOH$ parameter within the mixtures on the concrete slump is relatively minimal for ratios exceeding the threshold of 0 (2.0) at the same alkaline concentration. For instance, when the NaOH concentration is held constant at the +1 level (14 M) and the $Na_2SiO_3/NaOH$ ratio is elevated from the −1 level (1.5) to the +1 level (2.5), there is a notable maximum decrease in slump of approximately 40 mm.

The response surface graph presented in **Figure 12** substantiates the performance in a harmonic manner concerning the dependent variable (compressive strength) at 28 days. Overall, it was discerned through the independent variables that this enhancement is notably amplified for alkaline concentrations exceeding the 0 level (12 M) and $Na_2SiO_3/NaOH$ ratios greater than the 0 level (2.0) in conjunction, culminating in compressive strength values on the order of 41 MPa at 28 days.

Encoded variable	Equation*	F-Test	R^2
Final setting time	184.00−40.00 (1) + 7.50 (2) + 2.50 (3)	18.03	0.9772
Slump Test	97.20−5.75 (1)−21.00 (2) + 1.50 (3)	35.20	0.9883
Compressive strength: 28 days	33.06 + 0.83 (1) + 7.61 (2)−0.03 (3)	17.44	0.9764

**Equation: (1) NaOH; (2) Na2SiO3/NaOH; (3) NaOH. Na2SiO3/NaOH.*

Table 6.
Regression models for the responses of alkali-activated composites.

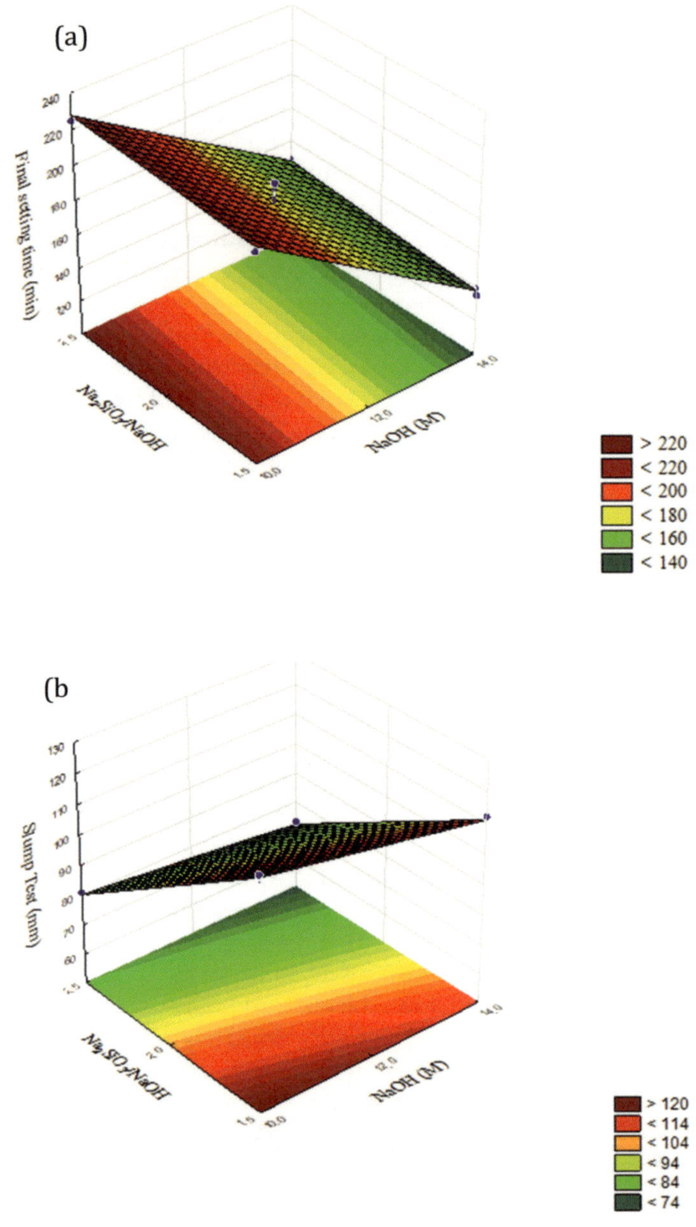

Figure 11.
Response surface for (a) final setting time and (b) slump.

In the pursuit of optimizing the dosing procedure for alkali-activated concrete, the preferred parameters outlined in **Table 7** were ascertained to uphold the requisite properties in both the fresh and hardened states. The objective function was established with an alkaline concentration of sodium hydroxide (NaOH) at 10 M and a ratio of sodium silicate to sodium hydroxide (Na_2SiO_3/NaOH) set at 2.5, while consistently preserving the same independent variable.

The optimization of the final setting time and the slump time of the concrete was performed based on Brazilian Association of Technical Standards [59]. The maximum

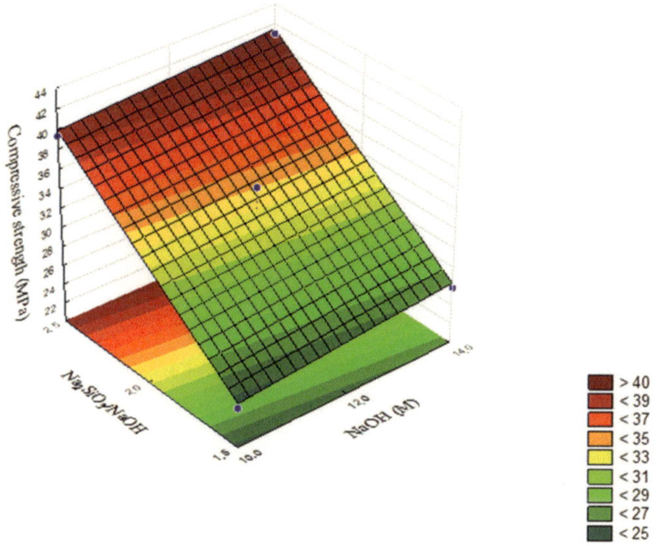

Figure 12.
Response surface for compression strength up to 28 days.

Response	NaOH	Na_2SiO_3/NaOH	Optimization
Final setting time	−1 (10 M)	+1 (2.5)	230 min
Slump	−1 (10 M)	+1 (2.5)	82 mm
Compressive strength: 28 days	−1 (10 M)	+1 (2.5)	39.50 MPa

Table 7.
Optimization of alkali-activated concretes.

time for complete application on site is 150 minutes, and a slump of more than 80 mm is required to obtain pumpable concrete. By setting the alkaline concentration up to 10 m and the Na_2SiO_3/NaOH ratio at 2.5, it is possible to obtain a viable mixture with a final setting time close to 230 minutes and a slump of approximately 82 mm.

In addition to fulfilling the properties in the fresh state, the optimization is directed toward meeting the criteria established in the Brazilian Association of Technical Standards [60] for structural concrete, which necessitates a compressive strength exceeding 20 MPa at 28 days. Utilizing an alkaline concentration of 10 M along with a Na_2SiO_3/NaOH ratio of 2.5, Figure 15 illustrates that it is feasible to achieve strengths surpassing 35 MPa at 28 days, taking into account that the model demonstrates statistical significance.

4. Conclusions

According to the results obtained, the following can be highlighted:

- Pumpable alkali-activated structural concretes (except T-12-3.0 and T-14-2.5) and capable of being machined (except T-14-1.5, T-14-2.5, and T-16-2.0) were obtained using a low-reactivity MKFeQ with high iron and quartz contents.

- *Fresh part:* Greater alkaline dissolutions accelerated setting and reduced slump, especially for higher $Na_2SiO_3/NaOH$.
- *Hardened part:* Higher combined alkaline concentrations of the highest ratio of alkaline reagents tend to increase mechanical strength.
- *Microstructure:* it was proven that not only is there an optimal concentration of alkaline concentration (NaOH) to guarantee the complete effectiveness of the dissolution of aluminosilicates, but also the increase in the ratio of alkaline reagents ($Na_2SiO_3/NaOH$) generates denser three-dimensional networks.
- *Dosage:* The optimized dosage indicated the lowest alkaline concentration (NaOH) (10 M) and the highest order $Na_2SiO_3/NaOH$ ratio (2.5).

Thanks

The authors would like to thank the Coordination for the Improvement of Higher Education Personnel - Brazil (CAPES) for the financial support granted and the Federal University of Paraiba (LABEME-UFPB) for the infrastructure granted for the project.

Author details

José Anselmo da Silva Neto[1*], Jhennifer Larissa de Barros Silva[2], José Augusto Gomes Neto[1] and Aline Figueirêdo da Nóbrega[2]

1 Federal University of Paraíba (UFPB), João Pessoa, PB, Brazil

2 Federal University of Campina Grande (UFCG), Campina Grande, PB, Brazil

*Address all correspondence to: neto.anselmo00@gmail.com

References

[1] Scrivener KL, John VM, Gartner EM. Eco-efficient cements: Potential economically viable solutions for a low-CO_2 cement-based materials industry. Cement and Concrete Research. 2018;**114**:2-26

[2] Caballero LR. Physical-Mechanical Behavior of Geopolymeric Matrices Based on Metakaolin Reinforced with Steel Fiber. Federal University of Rio de Janeiro; 2017

[3] Davidovits J. Geopolymer Chemistry and Applications. 4th ed. Saint-Quentin Fr: J. Davidovits; 2015

[4] Ishak S, Lee H-S, Singh JK, Ariffin MAM, Lim NHAS, Yang H-M. Performance of fly ash geopolymer concrete incorporating bamboo ash at elevated temperature. Materials (Basel). 2019;**12**:3404

[5] Silva MCD et al. Study of the mechanical performance of geopolymer produced with high iron content. Matter (Rio de Janeiro). 2023;**28**(1):e20220309

[6] Brazilian Association of Technical Standards (ABNT). NBR 15894-1—Metakaolin for Use with Portland Cement in Concrete, Mortar and Paste. Part 1: Requirements. Rio de Janeiro; 2010

[7] Lemougna PN, Mackenzie KJD, Jameson GNL, et al. The role of iron in the formation of inorganic polymers (geopolymers) from volcanic ash: A Fe Mössbauer spectroscopy study. Journal of Materials Science. 2013;**48**(15):5280-5286

[8] Kamseu E, Moungam LMB, Cannio M, Billong N, Chaysuwan D, Melo UC, et al. Substitution of sodium silicate with rice husk ash-NaOH solution in metakaolin based geopolymer cement concerning reduction in global warming. Journal of Cleaner Production. 2017;**142**:3050-3060 https://doi.org/doi:10.1016/j.jclepro.2016.10.164

[9] Singh K. Experimental study on metakolin and baggashe ash based geopolymer concrete. Materials Today Proceedings. 2020;**37**:3289-3295. DOI: 10.1016/j.matpr.2020.09.116

[10] Khalil WI, Frayyeh QJ, Abed HT. Properties of metakaolin based pervious geopolymer concrete. IOP Conference Series: Materials Science and Engineering. 2019;**518**:22056. DOI: 10.1088/1757-899x/518/2/022056

[11] Deepak AL, Lakshmi T. Mechanical properties of geopolymer concrete with flyash and metakaolin. International Journal of Engineering and Advanced Technology. 2019;**9**:3863-3868

[12] Reddy MS, Dinakar P, Rao BH. Mix design development of fly ash and ground granulated blast furnace slag based geopolymer concrete. Journal of Building Engineering. 2018;**20**:712-722. DOI: 10.1016/j.jobe.2018.09.010

[13] Nuruddin MF, Kusbiantoro A, Qazi S, Darmawan M, Husin NA. Development of geopolymer concrete with different curing conditions. IPTEK The Journal for Technology and Science. 2011;**22**:24-28. DOI: 10.12962/j20882033.v22i1.54

[14] Silva, Neto JA, Marçal NA, Nóbrega AF, Nóbrega ACV, Souza JJN, Malheiro R. Use of metakaolin with a low surface area and rich in quartz and iron as a precursor in the production of structural alkali-activated concrete. Construction and Building Materials. 2024;**136418**:1-12

[15] Ngnintendem DLV, Lampe M, Tchakouté HK, et al. Effects of iron minerals on the compressive strengths and microstructural properties of metakaolin-based geopolymer materials. Gels. 2022;**8**(8):525

[16] Gomes KC, Lima GST, Torres SM, Barros SR, Vasconcelos IF, Barbosa NP. Iron distribution in geopolymer with ferromagnetic rich precursor. Materials Science Forum. 2010;**643**:131-138

[17] Ahmed HU, Mohammed AA, Rafiq S, Mohammed AS, Mosavi A, Sor NH, et al. Compressive strength of sustainable geopolymer concrete composites: A state-of-the-art review. Sustainability. 2021;**13**:13502

[18] Moradikhou AB, Esparham A, Jamshidi Avanaki M. Physical and mechanical properties of fiber reinforced metakaolin-based geopolymer concrete. Construction and Building Materials. 2020;**251**:118965. DOI: 10.1016/j.conbuildmat.2020.118965

[19] Zhang HY, Qiu GH, Kodur V, Yuan ZS. Spalling behavior of metakaolin-fly ash based geopolymer concrete under elevated temperature exposure. Cement and Concrete Composites. 2020;**106**:103483. DOI: 10.1016/j.cemconcomp.2019.103483

[20] Thammarong S, Lertcumfu N, Jaita P, Pimpha N, Tunkasiri T, Rujijanagul G, Malasri P. Physical and Mechanical Properties of Diatomite/Metakaolin-Based Geopolymer for Construction Materials. 2020:1121-1148

[21] Alonso S, Palomo A. Calorimetric study of alkaline activation of calcium hydroxide—Metakaolin solid mixtures. Cement and Concrete Research. 2001;**31**:25-30. DOI: 10.1016/S0008-8846(00)00435-X

[22] Wang Y, Zheng T, Zheng X, Liu Y, Darkwa J, Zhou G. Thermomechanical and moisture absorption properties of fly ash-based lightweight geopolymer concrete reinforced by polypropylene fibers. Construction and Building Materials. 2020;**251**:118960. DOI: 10.1016/j.conbuildmat.2020.118960

[23] Lizcano M, Kim HS, Basu S, Radovic M. Mechanical properties of sodium and potassium activated metakaolin-based geopolymers. Journal of Materials Science. 2012;**47**:2607-2616. DOI: 10.1007/s10853-011-6085-4

[24] Liu X, Jiang J, Zhang H, Li M, Wu Y, Guo L, et al. Thermal stability and microstructure of metakaolin-based geopolymer blended with rice husk ash. Applied Clay Science. 2020;**196**:105769. DOI: 10.1016/j.clay.2020.105769

[25] Duxson P, Mallicoat SW, Lukey GC, Kriven WM, van Deventer JSJ. The effect of alkali and Si/Al ratio on the development of mechanical properties of metakaolin-based geopolymers. Colloids and Surfaces A: Physicochemical and Engineering Aspects. 2007;**292**:8-20. DOI: 10.1016/j.colsurfa.2006.05.044

[26] Allahverdi A, Vafaei M, Maghsoodloorad H. Quality control and assessment of geopolymer cements based on reacted and free alkalis. Construction and Building Materials. 2017;**153**:274-283. DOI: 10.1016/j.conbuildmat.2017.07.073

[27] Yuan J, He P, Jia D, Yang C, Zhang Y, Yan S, et al. Effect of curing temperature and SiO2/K2O molar ratio on the performance of metakaolin-based geopolymers. Ceramics International. 2016;**42**:16184-16190. DOI: 10.1016/j.ceramint.2016.07.139

[28] Catauro M, Dal Poggetto G, Sgarlata C, Vecchio Ciprioti S, Pacifico S, Leonelli C. Thermal and microbiological

performance of metakaolin-based geopolymers cement with waste glass. Applied Clay Science. 2020;**197**:105763. DOI: 10.1016/j.clay.2020.105763

[29] Zhang Z, Wang H, Yao X, Zhu Y. Effects of halloysite in kaolin on the formation and properties of geopolymers. Cement and Concrete Composites. 2012;**34**:709-715. DOI: 10.1016/j.cemconcomp. 2012.02.003

[30] Frayyeh QJ, Khalil WI, Abed HT. Sustainable metakaolin based pervious geopolymer concrete with recycled concrete aggregate. In: IOP Conference Series: Materials Science and Engineering. IOP Publishing; 2020. p. 12049

[31] Ahmed MF, Khalil WI, Frayyeh QJ. Thermal insulation enhancement of metakaolin-based geopolymer concrete using waste clay brick. IOP Conference Series: Materials Science and Engineering. 2020;**842**:12009. DOI: 10.1088/1757-899X/842/1/012009

[32] DeRousseau MA, Kasprzyk JR, Srubar WV. Computational design optimization of concrete mixtures: A review. Cement and Concrete Research. 2018;**109**:42-53. DOI: 10.1016/j. cemconres.2018.04.007

[33] Kashani A, Ngo TD, Walkley B, Mendis P. Thermal performance of calcium-rich alkali-activated materials: A microstructural and mechanical study. Construction and Building Materials. 2017;**153**:225-237. DOI: 10.1016/j. conbuildmat.2017.07.119

[34] Sarker PK, Kelly S, Yao Z. Effect of fire exposure on cracking, spalling and residual strength of fly ash geopolymer concrete. Materials and Design. 2014;**63**:584-592. DOI: 10.1016/j. matdes.2014.06.059

[35] Alghannam M, Albidah A, Abbas H, Al-Salloum Y. Influence of critical parameters of mix proportions on properties of MK-based geopolymer concrete. Arabian Journal for Science and Engineering. 2021;**46**:4399-4408. DOI: 10.1007/s13369-020-04970-0

[36] Askarian M, Tao Z, Adam G, Samali B. Mechanical properties of environment cured one-part hybrid OPC—Geopolymer concrete. Construction and Building Materials. 2018;**186**:330-337. DOI: 10.1016/j. conbuildmat.2018.07.160

[37] Brazilian Association of Technical Standards (ABNT). NBR NM 248: Aggregates–Determination of Particle Size Composition. Rio de Janeiro; 2003

[38] Zhang C, Hu Z, Zhu H, et al. Effects of silicate on reaction process and microstructure of metakaolin-based geopolymer composites. Journal of Building Engineering. 2020;**32**(June):101695

[39] Brazilian Association of Technical Standards (ABNT). NBR NM 52: Fine Aggregate—Determination of Specific Mass and Apparent Specific Mass. Rio de Janeiro; 2009

[40] Brazilian Association of Technical Standards (ABNT). NBR NM 53: Coarse Aggregate—Determination of Specific Mass, Apparent Specific Mass and Water Absorption. Rio de Janeiro; 2009

[41] Brazilian Association of Technical Standards (ABNT). NBR 16605: Portland Cement and Other Powdered Materials—Determination of Specific Gravity. Rio de Janeiro; 2017

[42] Brazilian Association of Technical Standards (ABNT). NBR 12653: Pozzolanic Materials—Requirements. Rio de Janeiro; 2014

[43] Silveira L, Acchar W, Gomes U, et al. Characterization of kaolin and granite residue for the formulation of porcelain stoneware. In: 22nd CBECiMat–Brazilian Congress of Engineering and Materials Science. Natal, RN, Brazil; 2016. pp. 1011-1022

[44] Zhang H et al. The effect of ordinary Portland cement substitution on the thermal stability of geopolymer concrete. Materials. 2019;**12**(16):2501-2521

[45] Melo. LGA Synthesis and Characterization of Geopolymers Containing Phyllites. 2011.184f. Dissertation (Master's in Materials Science). Rio de Janeiro: Military Institute of Engineering; 2011

[46] Perez-Cortes P, Escalante-Garcia J. Design and optimization of alkaline binders of limestone-metakaolin—A comparison of strength, microstructure and sustainability with Portland cement and geopolymers. Journal of Cleaner Production. 2020;**273**:123118-123221

[47] Brazilian Association of Technical Standards (ABNT). NBR 5738: Concrete—Procedure for Molding and Curing Specimens. Rio de Janeiro; 2015

[48] Brazilian Association of Technical Standards (ABNT). NBR 5739: Concrete—Compression Test of Cylindrical Specimens. Rio de Janeiro; 2018

[49] Brazilian Association of Technical Standards (ABNT). NBR 16607: Portland Cement—Determination of Setting Times. Rio de Janeiro; 2018

[50] Brazilian Association of Technical Standards (ABNT). NBR NM 67: Concrete—Determination of Consistency by Slumping the Frustum of the Cone. Rio de Janeiro; 1988

[51] Brazilian Association of Technical Standards (ABNT). NBR 7212: Concrete Dosed in a Plant—Preparation, Supply and Control. Rio de Janeiro; 2021

[52] Xie T et al. Mix design and mechanical properties of geopolymer and alkali activated concrete: Review of the state-of-the-art and the development of a new unified approach. Construction and Building Materials. 2020;**256**:119380

[53] Fang G, Ho W, Tu W, Zhang M. Workability and mechanical properties of alkali-activated fly ash-slag concrete cured at ambient temperature. Construction and Building Materials. 2018;**172**:476-487

[54] Herwani I, Pane I, Imran B. Compressive strength of fly ash-based geopolymer concrete with a variable of sodium hydroxide (NaOH) solution molarity. In: MATEC Web of Conferences, EDP Sciences. 2018. pp. 1-14

[55] Chithambaram S, Kumar S, Prasad M, Adak D. Effect of parameters on the compressive strength of fly ash based geopolymer concrete. Structural Concrete. 2018;**19**:1202-1209

[56] Usha S, Nair DV. Feasibility study of geopolymer binder from terracotta roof tile waste. Procedia Technology. 2016;**25**:186-193

[57] Brazilian Association of Technical Standards. NBR 8953: Concrete for Structural Purposes—Classification by Specific Mass, by Resistance and Consistency Groups. Rio de Janeiro; 2015

[58] Lizcano M et al. Mechanical properties of sodium and potassium activated metakaolin based geopolymers. Journal of Materials Science. 2018;**47**:2607-2616

[59] Brazilian Association of Technical Standards. NBR 7212: Concrete Dosed in a Plant—Prepared, Complete and Control. Rio de Janeiro; 2014

[60] Brazilian Association of Technical Standards. NBR 6118: Design of Concrete Structures. Rio de Janeiro; 2014

Chapter 2

Perspective Chapter: Geopolymers in Civil and Environmental Engineering Applications

Marianne Saba and Wahib Arairo

Abstract

This chapter highlights the potential applications of geopolymers in civil and environmental engineering applications. It indicated how geopolymeric materials could reform current building techniques and sustainable practices. This chapter will start by elaborating on and synthesizing geopolymer mortars, promoting the use of natural and recycled source materials with a focus on industrial waste and sustainable raw materials. This chapter delves into the characterization methods that enable geopolymers' good mechanical properties in the coming step. Among the significant properties are the correlation with the microstructural, thermal, and durability properties. The chapter also discusses how geopolymers help the environment by managing waste and reducing carbon footprints.

Keywords: geopolymer synthesis, characterization, sustainable materials, durability, mechanical properties

1. Introduction

This chapter will overview geopolymer mortars' synthesis, characterization, mechanical properties, and environmental benefits. Significant developments in past decades have been dedicated to developing more sustainable construction materials. There is a continuously rising demand for green concrete, emanating from the heightening demand for sustainable construction materials. Against this, this chapter looks explicitly into geopolymers, which fit into the class of inorganic polymers obtained from the reaction of aluminosilicate materials with alkaline activators. Geopolymeric material is sustainable because it has excellent mechanical properties and durability, particularly because of the great potential for reusing recyclable and industrial by-products and reducing carbon emissions. Finally, in this regard, the synthesis, characterization, and application of geopolymers in civil and environmental engineering are overviewed in this chapter of the present book, together with their relation to past and present synthesis.

The quest for more sustainable and durable construction materials has intensified interest in geopolymers, a class of inorganic polymers resulting from the alkali activation of aluminosilicate materials. Geopolymers offer a range of environmental and mechanical benefits when compared with traditional Portland cement; thus, it is quite a promising alternative to its application in civil and environmental engineering. Geopolymers are a particular case of materials known under the name of alkali-activated materials, which were forecasted as one of the possible applications as a green material for building [1].

Normally, geopolymers are alkali-activated aluminosilicates whose hydration products are presented by zeolite-like phases similar to natural zeolites [2]. They are the product of a reaction between aluminosilicate precursors and alkaline solutions when forming a three-dimensional polymeric network. Some of the most used precursors in the production of geopolymers are the by-products of industry, such as fly ash, metakaolin, and slag; hence, this material is labeled green due to the use of waste. Geopolymers have been foreseen over the last two decades as a reliable alternative to conventional Ordinary Portland Cement (OPC) materials. The global CO_2 emissions from the production of Portland cement are enormous, on the order of 8%, and most of these emissions are from the calcination step [3].

In contrast, geopolymerization occurs under ambient temperature and does not need energy-intensive calcination, which could contribute significantly to greenhouse gas (GHG) emissions. The geopolymers demonstrate superior mechanical characteristics, such as high compressive strength, excellent fire resistance, and durability under the attack of aggressive environments. Such is a consequence of the said good chemistry of the material changing due to good stability and excellent resistance to chemical attack. Accordingly, their application is diverse, and the thermal insulation and low shrinkage properties are good, making them applicable in several fields [4].

Geopolymers are amorphous to semicrystalline aluminosilicate gel on a microscale, giving them a highly mechanical-strong and chemical-resistant nature. The properties of the geopolymer material will be strongly dependent on and influenced by the nature of whole precursor materials and the conditions applied during the geopolymerization process, including the type and concentration of the alkaline activator [5]. In this regard, it has been experimentally illustrated that modifying these parameters can lead to the optimized performance of the mechanical properties for engineering applications. Geopolymers have found some applications in several civil engineering works, from the construction stage to infrastructure repair and rehabilitation [6]. Among several developed applications, one of the primary ones for geopolymers includes producing a green alternative to conventional concrete for structural purposes. It was seen that geopolymer structures are compatible with aggressive environments, mainly in marine and industrial sites. Geopolymers are also used in the *in-situ* remediation of hazardous waste by providing a methodology for stabilization that does not allow the mobility of heavy metals and other contaminants [7]. This is relevant to sustainable waste management practices because it eliminates environmental pollution and valorizes industrial by-products. Practical implications of the uses of geopolymers refer to the environmental sustainability engineering discipline, with its uses in water and wastewater treatments, soil stabilization, and encapsulation of radioactive waste. Geopolymer materials efficiently remove heavy metals and other pollutants from aqueous solutions since they possess large surface areas and availability of ion exchange capacity [7]. The essential advantage of the material in treating contaminated media is the opportunities afforded by tailoring to provide definite adsorption characteristics currently applied in a broad spectrum of environmental remediation. It was seen that geopolymer can be tested in the matrixes for the encapsulation of radioactive waste, as it immobilizes radionuclides without the possibility of leaching them into the environment. To this end, this is one area where geopolymers can assist in delivering human-informed solutions for resolving environmental issues and simultaneously offer a long-term issue to containment problems that some other materials may encounter. Even though features potentially valuable to geopolymer applications are a few, challenging traits limit their wide acceptance. Together, they constitute some critical sources of variability in standardized test methods, source materials, and mix design optimization for an intended application. Accordingly, the prospects of geopolymers in civil and environmental engineering will be very bright since the daily serious challenges posed by new applications that material science and engineering are being pressed forward to meet. Research is progressively filling these gaps in the weaknesses of geopolymer applications in the quest for a robust, cost-effective solution. Geopolymers relate to the upward trend of resilient, sustainable infrastructures, given the recent upsurge in sustainability interest attributed to the construction industry. In this respect, geopolymers have been available to the materials science of concretes as a support for developing sustainable and high-performance alternatives to traditional OPC. Geopolymers represent an upcoming civil and environmental engineering revolution based on properties associated with their versatile applications [6]. This chapter is intended to understand, from basic chemistry to practical applications, pointing toward a new material that can change the construction industry.

2. Geopolymer mortars synthesis

The simple principle of geopolymer synthesis lies in the geopolymerization reaction in which aluminosilicate precursors activated in an alkaline solution (usually sodium or potassium-based) form a three-dimensional polymeric network. That means it is a multicriteria process, and each stage depends on the type of raw material, concentration of alkaline activators, and curing conditions. In summary, one chooses appropriate aluminosilicate precursors, normally by-products of industry products such as fly ash slag, or naturally occurring such as kaolinite clay or metakaolin. The material chosen to be the precursor during synthesis will affect the properties of the geopolymer since the reactivity varies according to the concentration of the chemical elements available and their composition. The next step will be the preparation of the alkaline activator, usually a mixture of sodium or potassium hydroxide and silicates of sodium or potassium. Their type and amount will be essential, as they will govern the dissolution rate of the aluminosilicate materials and the subsequent polymerization process. The higher the activator concentration, the rawer material dissolution will occur, hence the higher completeness of the reaction and, consequently, the better mechanical properties of geopolymers [8].

On the other hand, a concentration that is too high may result in flash settings or the suspicion of workability. The alkaline activator reacts with the aluminosilicate precursor mentioned above, whereby it dissolves aluminosilicate particles, releasing silicate and aluminate species into the solution. These species react with each other through a series of condensation reactions, resulting in a gel-like network that further completes into a complex solid matrix. Again, creating this three-dimensional network gives geopolymers high compressive strength and hardness [9].

In most cases, the curing conditions play an essential role in developing the geopolymer structure. Curing can be realized under ambient temperatures or with the help of temperature acceleration. High curing temperatures generally enhance geopolymers' mechanical properties as they usually raise the degree of polymerization, reducing the time required to achieve ultimate strength. More significantly, controlled humidity and curing time are the parameters considered. Excess drying will consequently lead to cracking in geopolymer matrixes; it can be prevented by maintaining optimal moisture and imparting uniformity and strength. This very flexibility of design and performance underlines the potential of geopolymers for replacing traditional cementitious materials with a sustainable, high-performance alternative for modern construction needs. This would be possible, in that respect, with the exact tailoring of each step from the choice of raw materials to the curing conditions and finally to the optimized performance for geopolymers destined for specific uses in civil and environmental engineering applications [8, 9].

2.1 Raw materials

Materials selection is an essential process that would determine the resulting properties of the geopolymer. Basically, the materials can be classified into two types: natural and recycled. The first category includes kaolinite, metakaolin, volcanic ash, and other clay minerals with sufficient aluminosilicates. Metakaolin will be more favorable due to its reactivity derived from kaolinite heat treatment [10]. The second category includes recycled materials, like fly ash, ground granulated blast furnace slag, or red mud, mainly resulting as by-products from industrial activities. Fly ash, a by-product of coal combustion rich in silicon and aluminum presents a potential source for precursor synthesis in geopolymers. Ground granulated blast furnace slag (GGBFS) is a by-product of making iron and steel that contains calcium, which geopolymer can use to enhance its mechanical properties and setting time [10]. Another source is red mud, an industrial aluminum production waste rich in alkalinity and aluminosilicate content. Generally, it should be noted that the type of materials is selected according to the proper amount of Al_2O_3 (Fe_2O_3) available; as such, the selection of constituents is chosen accordingly.

The alkaline activators necessary for geopolymerization will dissolve aluminosilicate materials and react further with polycondensation. Common activators thus come from sodium hydroxide, potassium hydroxide, and sodium silicate. The concentration and the ratio of these activations set the reaction kinetics, the setting time, and the mechanical properties of the geopolymer. One important indicative factor is the molarity of the NaOH or KOH solution. Generally, the better the reaction grade and early-age strength, the higher the molarity will be. However, it is at the expense of probably providing an extremely short setting time and, therefore, a less workable mixture. Adding sodium silicate will increase the SiO_2/Al_2O_3 nanoparticle ratio and give a geopolymer binder with a binding property that imparts the needed self-strength and durability to the whole system [3, 4].

Mixing is a process by which dry aluminosilicate precursors are combined with the alkaline solution to obtain a homogeneous paste. In this process, the mixing time, sequence, and destination of equipment are all potentially in a position to impact the different aspects of the quality and uniformity of geopolymer mortar. Generally, mechanical mixing is preferred as it may offer an adequate distribution of components.

One of the most essential parameters concerning geopolymer technology's induction and development time, besides at room temperature, is curing conditions at higher temperatures relating to mechanical properties and microstructure. This could be done either at room temperature or by heating to an elevated temperature; it depends on the specific application requirement and the type of raw materials used. A rise of the curing temperature to 40–80°C and the fast-aging geopolymer derives its strength. For most applications, however, curing at ambient temperature alone would be adequate, mainly if GGBFS were used since it is known to have a latent hydraulic nature, and strength continues to be developed through the continued geopolymerization [10, 11].

3. Geopolymer characterization

The geopolymer shall be tested using a set of characterization techniques to determine its properties. These tests provide information about the microstructure, the chemical composition of a geopolymer, its thermal stability, and its durability, as they allow for predicting performance conditions for in-service situations.

3.1 Microstructural analysis

Geopolymers are aluminosilicate gels that are amorphous and mixed with crystallites, remaining unreacted particles, and pores. Since microstructural analysis is closely related to mechanical properties and durability, this will define the phases described above regarding their distribution and interaction [12].

Some applications study geopolymer matrix morphology and texture using scanning electron microscopy. Such images demonstrate the network of gels together with crystalline phases and material porosity. It will also be essential in determining the extent of compaction, particle packing, and distribution of the unreacted particle, which will help sufficiently assess the efficacy of synthesis processes. It helps to identify crystalline phases present in the geopolymer. While geopolymers are essentially amorphous, one can detect and estimate crystalline phases like quartz, mullite, or zeolites. X-ray diffraction (XRD) patterns provide information about the mineralogical composition and the precursors' degree of reaction. Considering the Fourier Transform Infrared (FTIR) analysis, this test investigates the chemical bonding and functional groups of geopolymer. Generally, FTIR spectra show bands related to Si-O, Al-O, and Si-O-Al bonds, proving the formation of an aluminosilicate network [10]. Thus, it would be logical to find a relationship between the variations of band intensities and positions and the degree of polymerization and nature of bonding in the geopolymer matrix.

3.2 Thermal properties

The excellent thermal properties of geopolymers make them extremely useful in high-temperature, resistance, or thermal insulation applications. Geopolymers are reported to show good thermal stability, enabling them to remain structurally intact up to 1000°C [13]. This property is beneficial in civil engineering, especially for applications like fireproofing, thermal insulation, and high-temperature industrial processes where conventional materials lose integrity or deteriorate.

Differential scanning calorimetry (DSC) and thermogravimetric analyses can be used to establish thermal properties in geopolymers. DSC measures heat flow that attends phase transitions, providing critical information about thermal stability and composition. This helps in understanding geopolymers' thermal behaviors under varied temperature conditions, which may be essential in specific applications where such materials are put through cyclic thermal loads. On the other hand, TGA measures a material's weight change as a function of temperature. The technique determines the geopolymer's thermal degradation and volatile ingredients. Based on weight loss patterns, TGA can determine thermal stability and establish the makeup's nature of a material's composition. One of the essential parameters concerning geopolymers in thermal applications includes thermal conductivity. Most geopolymers usually have low thermal conductivity due to their amorphous nature and high porosity. Hence, they are very suitable for thermal insulation. For instance, geopolymers could be applied in the building and construction industries to make fire-resistant panels and insulation materials to improve safety and energy efficiency within a building. The very low thermal conductivity guarantees minimal heat transfer, hence effective insulation, which reduces energy consumption for heating and cooling purposes.

High-temperature resistance also qualifies geopolymers for refractory applications. These geopolymer-based products can line furnaces, kilns, or any other high-temperature equipment involved in industrial processes. The high-temperature resistance with less degradation adds to the service life of the installations and thus reduces maintenance costs. Besides, these fire-resistant properties can extend the application of geopolymers to passive fire protection systems for buildings or infrastructure. In a way, geopolymers can act as an intumescent coating or barrier, delay the fire, and provide the critical time required for evacuation and rescue operations in case of a fire outbreak.

Their thermal stability also benefits them in thermal cycling applications when the materials are subjected to repeated heating and cooling. In such conditions, traditional materials experience thermal fatigue and hence develop cracks. However, geopolymers do not have issues related to robust thermal properties, which ensure that they are long-lasting and durable, thus fitting within environments like power plants, aerospace, and the automotive industry where materials experience extreme fluctuations in temperature [14].

3.3 Durability

Material durability is a prime concern for their long-term performance and structural integrity in civil engineering. Blended geopolymers have proved superior performance over Portland cement-based concrete materials for all sorts of tests, mainly due to the dense microstructure that renders them less vulnerable to chemical attacks than the amorphous matrix classically observed for Portland cement, as stated by Provis et al. [8]. This intrinsic chemical resistance makes geopolymers more resistant to attack by acids and sulfates, thus allowing them to be used in aggressive environments. Compared to Portland cement, the reduced calcium content in geopolymer structures enables such materials to increase their resistance to sulfate attack and related deterioration mechanisms.

One of the common properties of geopolymers in cold regions is their resistance to freeze-thaw. In geopolymer-based materials, due to lower porosity and higher density, frost damage is prevented, which occurs due to the expansion of water within the pores in any material during a freezing and thawing cycle. This would be important in ensuring that materials have good structural integrity under adverse climatic conditions during cold weather, where traditional materials might be affected by repeated freeze-thaw cycles [14].

Also, geopolymers show lower chloride ion permeability, which is very important in reducing the corrosion of reinforcement concrete structures. Among the significant causes of deterioration of reinforced concrete, chloride-induced corrosion is condensed because of reduced permeability in geopolymers. In addition, geopolymers are highly resistant to carbonation. This is another degradation process common in concrete structures under the action of CO_2. It may be that such carbonation resistance would increase the lifetime of concrete structures, making geopolymers a feasible and long-lasting alternative to traditional materials [15].

Given improved durability and resistance to chemicals, besides freeze-thaw and chloride, geopolymers offer massive potential for application in civil engineering, where long-term performance and sustainability are prime essentials. Because of these characteristics, geopolymers have a wide range of usages, especially in infrastructure works in marine and industrial environments, generally exposed to harsh chemicals and climatic conditions.

3.4 Compressive strength

One of the fundamental mechanical properties related to civil engineering that shows a material's ability to withstand loads along its axis without failure is compressive strength. Geopolymers generally show quite impressive compressive strengths, mostly above 60 MPa. This strength is partly due to the formation of an aluminosilicate gel network that is very dense and continues to hold the aggregates and fillers firmly together. Factors such as the precursor material's nature, concentration, the alkaline activator's nature, and curing conditions may influence compressive strengths in geopolymers. Ground granulated blast furnace slag is one of the most commonly added to a geopolymer mix to increase its compressive strength. GGBFS participates in forming calcium-aluminate-silicate-hydrate phases similar to the calcium-silicate-hydrate phases formed in Portland cement. It is also known that these C-A-S-H phases improve the binding and enhance the geopolymer matrix's overall compressive strength. In a work conducted by Saba et al., different geopolymer mixes with varying proportions of metakaolin were tested [16]. The results showed that compressive strengths reached up to 80 MPa at 28 days for specimens cured at ambient temperature.

3.5 Flexural strength and toughness

Flexural strength, otherwise known as the modulus of rupture, is normally a measure of whether the material can withstand deformation by bending. Normally, geopolymers indicate large flexural strengths that recommend them for tension applications. Adding reinforcement fibers can improve this flexural performance: steel, glass, and polymer fibers work to bridge cracks and distribute stresses better to withstand bending forces more effectively. In a study by Rashad, the flexural strength of fiber-reinforced geopolymers was studied [17]. It was seen that the addition of steel fibers to a volume fraction of 0.015 showed an improvement in flexural strength from 5 to 8 MPa. In the same way, polymer fibers improve the toughness of geopolymer composites so that they can absorb more energy before failure. This gain in strength is of particular interest for structural applications where materials undergo dynamic loading and impact conditions.

3.6 Modulus of elasticity

The modulus of elasticity, commonly known as Young's modulus, is a measure of a material's stiffness and, thus, a predictor of its ability to withstand elastic deformation under stress. Geopolymers generally have lower elasticity moduli than Portland cement concretes, which becomes an advantage in applications demanding ductility or capacity to accommodate deformation. In most cases, the modulus of elasticity of geopolymers is generally governed by precursor types, alkaline activator concentration, and curing conditions. For example, Saba et al. demonstrated that geopolymers synthesized with metakaolin and sodium hydroxide activators showed modulus of elasticity values in the 10–20 GPa range [18]. The lower modulus of elasticity can be useful in enhancing the energy absorption/dissipation capacity during failure and reduces the propensity for brittle failure. Lower stiffness makes geopolymers suitable for seismic-resistant structures, requiring materials to bear high strains without losing integrity.

4. Environmental advantages

Geopolymers, undoubtedly, have the most potential for being green materials for sustainable construction in terms of environmental performance. Among the significant benefits of geopolymers is that they offer an immense opportunity for mitigating the environmental impact caused by conventional Portland cement, responsible for about 8% of global CO_2 emissions. Concerning this, geopolymers considerably decrease these emissions and offer a great substitute for civil works while taking care of sustainability [19].

4.1 Waste management

The application of numerous types of industrial wastes, such as fly ash, slag, red mud, and so on, in geopolymers, testifies very well to a well-thought-out mechanism of waste management on the part of the construction industry. Since these by-products are generally considered wastes, they can effectively be used in geopolymer formulations to divert materials from landfills and reduce environmental degradation. For instance, fly ash is one of the by-products of coal combustion, which enhances the mechanical properties of geopolymers and reduces the volume of virgin material used, hence saving natural resources. The valorization of industrial waste in the production of geopolymers works twofold: it solves disposal problems because there is a great directive in the European Union to regulate how waste is disposed of while creating an economic incentive through reduced raw material costs. This valorization route promotes a circular economy for geopolymers, in which the waste is a precious construction product that, in turn, represents resource efficiency and sustainability in the different civil engineering practices [20]. Up to 70% of industrial by-products can be shown to be utilized in geopolymer formulations, with the materials staying high-performance, significantly reducing the ecological footprint of construction activities [7].

4.2 Carbon footprint reduction

The carbon footprint from geopolymer production is low compared with ordinary Portland cement for some key reasons. First, the geopolymerization reaction occurs at relatively low temperatures, normally 40–100°C, instead of cement clinker production at 1450°C. Processing at these lower temperatures reduces energy consumption, hence decreasing the general carbon emissions associated with geopolymer production. This means that most geopolymers' raw materials are directly obtained from waste streams, while the latter are generally disposed of in an environmentally damaging manner. Using such materials, geopolymers reduce natural resource use and lower CO_2 emissions by avoiding extensive quarrying and processing of raw materials. According to studies within a life cycle assessment framework, geopolymers reportedly reduce CO_2 emissions by more than 80% compared with traditional Portland cement applications, making them a likely sustainable alternative for civil engineering [5]. Indeed, the dense microstructure is formed mainly by creating a three-dimensional aluminosilicate network, which confers improved resistance to chemical attack, freeze-thaw cycles, and other forms of environmental stress. Increased durability extends the service life of structures and reduces maintenance costs and consumption of resources over time. In other words, if geopolymers are added to different types of infrastructure or civil works, long-term savings will be realized, and the environmental impacts associated with replacing and repairing a material will be reduced [21]. The application fields of geopolymers in infrastructure works, roads, bridges, and building applications fully meet modern green building initiatives and goals of sustainable urban development. In this regard, one shall assist in constructing resilient and sustainable infrastructures that answer the increasingly stringent environmental regulations and standards by opting for geopolymer-based materials in civil engineering works.

5. Applications of geopolymers in civil engineering

Geopolymers have limited applications in environmental engineering. They could be used in soil stabilization, waste containment, and water treatment. The geopolymer binders can immobilize the contaminated soil by effectively anchoring the heavy metal through fixation, reducing its leachability and preventing further environmental contamination [7]. This attribute is very essential in eliminating common sites for environmental feuds. Apart from the goal of soil stabilization, geopolymers provide an excellent source of confinement for hazardous wastes. In this respect, the geopolymerization steps end up with a stable structure that can bear toxicity for long periods and is encapsulated with impermeable barriers, confining the waste to provide environmental safety and satisfy the set regulatory standards. Further, continued research in applying geopolymers for water treatment has established that they can separate heavy metals and other wastes from wastewater, hence providing cleaner water resources [7].

Moreover, geopolymers have progressively attracted much attention in civil engineering due to their perfect subsistence for other conventional materials, wherein paste use and binders find applications in renovations, pavement constructions, infrastructural developments, and environmental engineering solutions. Geopolymers can be used as very effective pastes in many construction applications. High viscosity and excellent adhesion allow them to fill the gaps, build up the surface, and form seamless joints of the construction elements.

5.1 Infrastructure

Geopolymers have been gaining increasing interest in civil engineering due to their versatility and contribution to a cleaner environment. Areas where geopolymers become adequate substitutes for conventional materials include renovation binders, pavement construction, infrastructure development, and environmental engineering solutions [15].

Nowadays, geopolymers find increasingly growing applications in all types of infrastructure, moving beyond pavements and bridges to various civil structures. They have high compressive strengths, developed by the geopolymerization process, and can be used for load-bearing applications. Geopolymers represent high-performance materials that perform very well in the most aggressive conditions against chemical attacks and cycles, thus indicating excellent durability of infrastructure materials. Geopolymer concrete was selected for pavement applications because of its high compressive strength and durability. Excellent performance under cyclic loading and heavy traffic loading without structural damage places geopolymer concretes among those more sustainable options for modern roadways. Coupled with lower carbon emissions and the use of industry by-products associated with geopolymers, they are in line to satisfy the increasing demand for green construction materials within the sectors of infrastructure projects. In building construction, geopolymers may replace traditional concrete and masonry materials. Additionally, geopolymers can be easily used in wall insulation and their fireproofing applications due to their low thermal conductivity and the fact that they are resistant to fire, improving the strength and safety against fires and the energy consumption of structures. As per Assaad et al., geopolymer-based materials can also be used with other precast elements, including panels and blocks, offering high strength and durability and the benefits of rapid construction [22].

5.2 Binder for renovation

One of the promising fields of application for geopolymers is as a binder in renovation works. If geopolymers are materials with excellent adhesion properties and can bond well with existing substrates, they can very well substitute conventional cement-based materials for repair and restoration. Therefore, due to the high resistance against chemical attacks, high compression strength, and durability, they will fit for retrofitting in historical structures, wherein keeping structural integrity is paramount while keeping contextual intervention on esthetics to a minimum.

Geopolymer binders have the potential to improve durability and repair performance significantly, which results from strong bonding and resistance to environmental attacks by water and contaminants. Adding further performance improvers into geopolymer binders, such as recycled aggregates and wastes, to their formulation can enhance their performance and develop an economically feasible and environmentally friendly solution for renovation works [23].

Moreover, working flexibility in geopolymer formulations allows the design to be tailored to the requirements of the projects, making it suitable for renovation works, even for residential buildings, or the vast improvement of infrastructure on a large scale. The application for infrastructure works now uses geopolymers in an expanding manner, covering beyond pavements and bridges to a wide range of other types of civil structures. This geopolymerization can offer high compressive strengths so that different materials can be successfully mounted in load-carrying applications. These geopolymers exhibit excellent behaviors under aggressive conditions, resisting chemical attacks and freeze-thaw cycles, two of the most critical parameters governing durability in infrastructure.

High compressive strength and immense durability are dazzling treats that make geopolymer concrete a lucid option for pavement construction. Geopolymer can manage huge traffic loads without experiencing any structural failure, making it a material for modern roads with a low carbon footprint. Further, the environmental benefits of geopolymers due to reduced carbon emissions and utilization of industrial by-products have recently come into the limelight, indicating a growing demand for green construction materials in infrastructure projects [24].

As substitutes to traditional concrete and masonry, geopolymers can offer an alternative solution to making building construction more sustainable. This is more useful in wall insulation and fireproofing since it has low thermal conductivity and raises building usage's safety by being fire-resistant, which improves energy efficiency. According to Assaad et al., geopolymer-based materials are expected to be used in high strength and durability precast elements such as panels and blocks for rapid construction [22].

The addition of fibers helps bridge micro-cracks that might develop under loading, making them distribute the stress efficiently and thus averting any crack propagation. Some studies have indicated that incorporating fibers into geopolymer composite results in improved ductility and resistance to impact loading, making it suitable for use when a material is subjected to transmit loads arising from dynamic forces. This trait can be helpful in structural members under seismic activity or heavy loads, as a gain in toughness would ease sudden failure.

It was proven that fiber-reinforced geopolymers have an advantage over ordinary concrete in terms of resistance to fire. The characteristic nature of such composites, bearing high temperatures, makes them very suitable for fire-prone environments connected with the non-destruction of structure. The fibrous reinforcement contributed to the whole durability that would retain the mechanical properties of that material under extreme conditions. Geopolymers are, therefore, new-generation materials that can potentially revolutionize the field of civil engineering through diversified applications in several industries [4]. Their functionalities, such as paste, renovation binder, and durable solution for infrastructure development, enable them to become a sustainable solution for modern challenges in construction. Consequently, geopolymers will soon find wide applications in civil engineering, offering innovative and sustainable solutions for an industry under increasing pressure concerning using more environmentally friendly construction materials and efficient constructability [5].

6. Conclusion

It can be concluded that geopolymers represent a new frontier in the application of material science to civil and environmental engineering. Their properties make them suitable, if not excellent, alternatives to cement-based materials: their high compressive strengths, durability, and improved resistance to most environmental stresses. The synthesis of geopolymers from by-products of industry fly ash, slag, and metakaolin reduces construction's carbon footprint and implements sustainable waste management practices. Fiber reinforcement further ameliorates the mechanical properties of geopolymer composites, allowing them to find a place in a broad scope of applications, spanning from pavement and bridge building to insulation and fire-resistant ones.

Critical advantages for infrastructure applications lie in the excellent resistance of geopolymers against the most aggressive environmental conditions due to freeze-thaw cycles or chemical attacks, and their low thermal conductivity and exceptional fire resistance are all valuable characteristics for constructing more efficient buildings that offer high performance and sustainability. Their versatility spans from paste formulation to binders for renovation, soil stabilization, and hazardous waste encapsulation, which are linked to civil engineering applications. Without a doubt, the role of geopolymers will further increase as the construction industry moves ever more toward embracing sustainability. New precursors and activation methods, and more importantly, new applications, continue to be driven by industry, further improving the performance and sustainability of geopolymer-based materials. Further studies are needed on mix design optimization, long-term durability under actual exposure conditions, and the establishment of comprehensive guidelines for the effective use of geopolymers in civil engineering works.

Ultimately, geopolymers stand symbolically for a solution that will outlast current environmental challenges and the pathway to the kind of future construction materials needed to realize performance, resilience, and ecological responsibility. Given such enormous challenges as climate change and resource depletion for resilient built environments, broadening adoption and further development of geopolymers should henceforth form part of the sustainable development goals.

Author details

Marianne Saba[1*] and Wahib Arairo[2]

1 Faculty of Engineering, University of Balamand, Lebanon

2 Issam Fares Faculty of Technology, University of Balamand, Lebanon

*Address all correspondence to: marianne.saba@balamand.edu.lb

References

[1] Provis JL, Van Deventer JSJ, editors. Alkali Activated Materials: State-of-the-Art Report, RILEM TC 224-AAM. Vol. 13. Dordrecht, Heidelberg, New York, London: Springer Science & Business Media; 2013

[2] Duxson P, Fernández-Jiménez A, Provis JL, Lukey GC, Palomo A, van Deventer JS. Geopolymer technology: The current state of the art. Journal of Materials Science. 2007;**42**:2917-2933

[3] Davidovits J. Geopolymer Chemistry and Applications. Morrisville, NC, USA: Geopolymer Institute; 2008

[4] Cong P, Cheng Y. Advances in geopolymer materials: A comprehensive review. Journal of Traffic and Transportation Engineering (English Edition). 2021;**8**(3):283-314

[5] Singh B, Ishwarya G, Gupta M, Bhattacharyya SK. Geopolymer concrete: A review of some recent developments. Construction and Building Materials. 2015;**85**:78-90

[6] Almutairi AL, Tayeh BA, Adesina A, Isleem HF, Zeyad AM. Potential applications of geopolymer concrete in construction: A review. Case Studies in Construction Materials. 2021;**15**:e00733

[7] Maleki A, Hajizadeh Z, Sharifi V, Emdadi Z. A green, porous and eco-friendly magnetic geopolymer adsorbent for heavy metals removal from aqueous solutions. Journal of Cleaner Production. 2019;**215**:1233-1245

[8] Provis JL, Van Deventer JSJ, editors. Geopolymers: Structures, Processing, Properties and Industrial Applications. Cornwell, UK: Elsevier; 2009

[9] Wu Y, Lu B, Bai T, Wang H, Du F, Zhang Y, et al. Geopolymer, green alkali activated cementitious material: Synthesis, applications and challenges. Construction and Building Materials. 2019;**224**:930-949

[10] Part WK, Ramli M, Cheah CB. An overview on the influence of various factors on the properties of geopolymer concrete derived from industrial by-products. Construction and Building Materials. 2015;**77**:370-395

[11] Morsy AM, Ragheb AM, Shalan AH, Mohamed OH. Mechanical characteristics of GGBFS/FA-based geopolymer concrete and its environmental impact. Practice Periodical on Structural Design and Construction. 2022;**27**(2):04022017

[12] Shilar FA, Ganachari SV, Patil VB, Khan TY, Khadar SDA. Molarity activity effect on mechanical and microstructure properties of geopolymer concrete: A review. Case Studies in Construction Materials. 2022;**16**:e01014

[13] Ali MF, Natrajan MV. A review of geopolymer composite thermal properties. In: IOP Conference Series: Earth and Environmental Science. Vol. 822, No. 1. Montreal: IOP Publishing; 2021. p. 012051

[14] Wong LS. Durability performance of geopolymer concrete: A review. Polymers. 2022;**14**(5):868

[15] ElKhatib L, Al Aridi F, ElKordi A, Khatib J. Mechanical and durability properties of geopolymer concrete–a review. BAU Journal-Science and Technology. 2022;**3**(2):8

[16] Saba M, Tehrani FF, Hajikarimi P, Absi J. Investigation on partially replacing metakaolin with marine shell

waste to produce sustainable eco-friendly geopolymer mortars. Clean Technologies and Environmental Policy. 2023;**25**(8):2639-2653

[17] Rashad AM. Effect of steel fibers on geopolymer properties–the best synopsis for civil engineer. Construction and Building Materials. 2020;**246**:118534

[18] Saba M, Fakhari-Tehrani F, Michaud P, Hajikarimi P, Absi J. Experimental and numerical investigation of sodium-and potassium-based alkali activator on the mechanical properties of geopolymer-mortars using Lebanese kaolin. International Journal of Civil Engineering. 2021;**19**:1007-1020

[19] Rangan BV. Geopolymer concrete for environmental protection. The Indian Concrete Journal. 2014;**88**(4):41-59

[20] Farooq F, Jin X, Javed MF, Akbar A, Shah MI, Aslam F, et al. Geopolymer concrete as sustainable material: A state of the art review. Construction and Building Materials. 2021;**306**:124762

[21] Parathi S, Nagarajan P, Pallikkara SA. Ecofriendly geopolymer concrete: A comprehensive review. Clean Technologies and Environmental Policy. 2021;**23**:1701-1713

[22] Assaad JJ, Saba M. Use of Metakaolin and slag geopolymer adhesives for fixing tiles. ACI Materials Journal. 2024;**121**(4):3-16

[23] Saba M, Arairo W, Sabalbal G, El Bachawati M, Absi J. Use of aluminum powder for the development of a sustainable paste used in the restoration of historical monuments. In: E3S Web of Conferences. Vol. 436. France: EDP Sciences; 2023. p. 08001

[24] Bernal SA, Provis JL. Durability of alkali-activated materials: Progress and perspectives. Journal of the American Ceramic Society. 2014;**97**(4):997-1008

Chapter 3

Performance Enhancement of Geopolymers through Natural Fiber Reinforcement: Sustainable Alternative Construction Materials

Zenamarkos Bantie and Desalegn Abera

Abstract

Cement manufacturing consumes a huge amount of energy and raw materials, resulting in the emission of greenhouse gases responsible for global warming, resource depletion, and environmental degradation. Recently, geopolymer composites emerged as promising alternatives to the traditional cementitious materials for their cost-effectiveness and environmental-friendly processing technology, offering similar technical properties as traditional cements with superior performances. However, geopolymers have relatively low tensile and flexural strengths, fragile breakage, and low ductility, limiting their use in the construction sector. Recent advances in the area revealed that fiber-reinforcing geopolymeric matrices overcome these and other related deficiencies. Various natural fibers have been researched as reinforcement of geopolymer matrices to provide improved tensile and flexural strengths, reduced density, and improved thermal and acoustic insulation properties. This review is intended to analyze and discuss the performance enhancements and macro- and microstructural modifications of geopolymer composites through natural fiber reinforcement.

Keywords: geopolymers, reinforcement, natural fibers, performance enhancement, sustainable material, durability

1. Introduction

The rabid growth of urbanization leads to the huge consumption of Portland cement, which causes a large amount of energy consumption and adversely affects the environment. Searching for a sustainable building material is the primary focus in the construction industry, which encourages researchers to develop alternative supplementary cementitious materials that partially or fully replace the cement. One emerging alternative technology in the construction industry is the usage of Geopolymer Concrete (GPC) [1, 2], which is synthesized from geological materials found in natural sources and some other industrial wastes. The introduction of GPC technology leads to sustainable development in the construction sector, reducing the global emission of CO_2 and providing a cleaner environment in the long term [3–5]. Nevertheless, these novel alternative cementing materials share similar shortcomings with hardened cement materials, including brittleness and comparatively low tensile strength [6, 7]. This problem could be tackled by incorporating fibers into the base aluminosilicate materials [8–10]. Unlike typical brittle materials where cracks spread quickly and significantly reduce the matrix's ability to bear loads, Fiber-Reinforced Concrete (FRC) behaves differently [11]. The fibers dispersed throughout the concrete matrix in FRC can impede crack growth, effectively slowing or even halting the formation of cracks.

Reinforcing geopolymer has been practiced with different fibrous materials [12–14]. Synthetic fibers, including steel, glass, and carbon fibers, are the most frequently utilized materials for reinforcing geopolymer [15–17]. Additionally, numerous studies have been conducted to examine the application of polymeric fibers, such as polypropylene and polyvinyl alcohol, as reinforcements of geopolymers [6, 11]. Li et al. [11] investigated the reinforcement of coal gangue-slag-based geopolymer with polypropylene fiber. The fiber significantly improves the flexural strength and reduces the brittle failure of the geopolymer. A steel fiber was utilized as reinforcement in beams of geopolymer composite [16] to modify the flexural properties of the composite. These synthetic-based fibers are based on limited fossil sources and cause an environmental problem [13]; thus, natural fiber would be the alternative for them.

Natural fibers, mostly available as waste materials, which may pose environmental problems, can be utilized as sustainable reinforcements of construction materials [18, 19]. Agricultural wastes and plant residues have the potential to be a source of natural fibers with comparable performance to synthetic fibers [10, 20]. Several scientific studies have examined the impact of incorporating various quantities and varieties of natural fibers on the properties of geopolymer composites [8, 14, 20]. The type, length, alignment, surface properties, and pre-treatments of these fibers also have an impact on the mechanical property improvement of the geopolymer. Zulfiati et al. [14] utilized pineapple leaf fiber for the modification of fly ash-based geopolymer, and the result revealed that as the length of fiber increased, the mechanical strength of the geopolymer also increased. Another study [2] evaluated South African alien invasive wood and sugarcane bagasse residues for the construction of panels used in outdoor conditions. The products exhibit comparable characteristics to other composites reinforced with natural fibers and can serve as alternative materials in similar applications. Silva et al. [8] studied the natural fibers, jute, and sisal for the performance enhancements of geopolymer composites from fired clay brick powder. The study examined the effect of the type and contents of fibers on the mechanical properties of geopolymer composites. Jute-fiber-reinforced geopolymer composites with 1.5% fiber content showed the highest increases in compressive and tensile strengths (64 and 45%), while sisal-FRGCs performed best at 2.5% fiber content. This chapter covers the general overview of natural fiber-reinforced geopolymers. Not inclusively all research published with natural fiber-reinforced geopolymers is reviewed. The synthesis, characterization, and performance of natural fiber-reinforced geopolymer were discussed in detail with current findings. Additionally, the future perspective of natural fiber-reinforced geopolymer was also highlighted.

2. Characterization of natural fibers

Natural fibers are gaining attention as environmentally friendly alternatives to synthetic materials in various applications such as textile and apparel, construction and building, automotive, arts and crafts, packaging, agriculture, etc. industries [21–25]. In this section, we aimed to discuss specifically the characterization of natural fiber to be applied in reinforcements of geopolymer composite. When characterizing natural fibers for use as reinforcements in geopolymers, multiple approaches are required to ensure that the fibers operate efficiently within the composite material. For this application, characterizing the physical, mechanical, chemical, fire resistance, thermal, moisture absorbance, etc. properties are required.

The physical properties of natural fiber to be considered include density, moisture content, ash content, diameter, and length measurements. Researchers have investigated the characterization of these physical properties with different fiber sources and standard methods [21, 25–27]. Geremew et al. [25] evaluated the bulk density (pycnometer), particle density, average diameter, moisture content (ASTM D 2216), ash content (ASTM D 2866), and water absorption of alkali-treated bamboo fiber. They obtained 0.065 g/cm^3, 0.685 g/cm^3, 1.158 mm, 7.690, 1.140, and 102.243%, respectively. Moussaoui et al. [28] determined the density of the *Ampelodesmos Mauritanicus* (AM) fibers using a pycnometer. AM fibers have a 1.00 g/cm^3 density, making them a suitable substitute for synthetic fibers in lightweight applications. The diameters of roselle fibers were measured with a Zeiss optical microscope [29]. The study evaluated the effect of growing moths with diameter and found that as months increase from 3, 6, and 9, the diameter range also increases from 40 to 100, 80 to 120, and 90 to 150 μm, respectively.

The chemical compositions of fibers also have an effect on their performance. Since the source of fiber is lignocellulosic material, extraction to its fraction is mandatory. Alessandro J. [30] extracted the lignocellulosic components of bamboo fibers determined according to NBR 14660. The bamboo fiber was extracted in toluene, alcohol, and water in the sequence of solvents. The lignin content was determined with the help of the spectrophotometer FEMTO 700 S, where the fibers were hydrolyzed with 72% sulfuric acid. Varma VK. and Sarangi SK. [31] analyzed the chemical configuration of the raw *Bauhinia Variegata* (BV) fiber (α-cellulose, hemicellulose, lignin, and wax), which was subjected to aging with 95%. Ethanol. The Technical Association of the Pulp and Paper Industry (TAPPI) standards and the T-203 method were used to measure the α-cellulose and hemicellulose content [29]. The Klason lignin technique to determine the lignin content of BV was employed.

Geopolymers are inorganic polymers that can benefit from mechanical reinforcement from natural fibers, which increases their strength, durability, and overall performance [22, 28]. The mechanical properties like tensile strength, elongation at break, and elastic modulus are basic to characterize. Moussaoui N. et al. [28] characterize the mechanical performance of the microwave, NaOH, and acetic anhydride-treated AM fibers through stress, elongation at failure, and modulus of elasticity with 40 mm gauge length. This study found that microwave irradiation with a power of 550 W for 4 minutes is the most effective way to change AM fiber surfaces. It increases strength from 155.41 to 289.93 MPa (86.55% increase) and Young's modulus from 10.99 to 18.59 GPa (69.15% increase). Another study [31] also investigated the mechanical properties of BV natural fiber treated with bleaching, benzoylation, and alkalization with a gauge length of 200 mm. Alkalized BVR fiber has a high tensile strength (50.52 MPa), Young's modulus (31.3 GPa), and 5.9% elongation, making it a viable alternative to traditional fibers in green composite manufacture. Hence, the mechanical properties of fibers are influenced by the nature of the fiber source, type of treatment used, length and diameters of fibers, standard methods used, etc. For the application of reinforcement, the mechanical performance of fiber is very important to impart this value to the geopolymer matrix.

Other structural and morphological characterizations are also important to tell us the significance of the fiber for the desired application. FTIR describes the interactions of chemical components of the fiber with the main matrix of material to be reinforced [27, 31]. The morphological characteristics of the fiber are also determined by Scanning Electron Microscopy (SEM) [27, 28, 30, 32] and Transmission Electron Microscopy (TEM) [33]. TEM visualizes the fiber's internal structure, including arrangement, orientation, hemicellulose, and lignin, aiding in understanding the relationship between the fiber's mechanical, thermal, and chemical properties. The SEM technique visualizes fiber surfaces, defects, and the fiber-matrix interface in composite materials, providing valuable insights into the fiber's processing, treatment, and performance. It also aids in investigating the effect of morphology on the material's mechanical properties. The crystalline structures of fiber are determined using an X-ray diffractometer [22, 25, 33]. The XRD analysis measures the crystallite structure of materials, such as crystallinity index and crystal size. Thermal properties of natural fibers can be characterized via a thermal conductivity meter [23] and Thermogravimetric Analysis (TGA) [22, 34]. The higher thermal resistance performance of fibers would guarantee their incorporation into the main matrix of material and will not pose drawbacks to the modified materials.

3. Synthesis of natural fiber-reinforced geopolymer composites

Natural fiber-reinforced geopolymer composites are a fascinating field of materials research that combines the environmental benefits of natural fibers with the distinct features of geopolymers. Geopolymer materials, such as cementitious matrices, exhibit strong quasi-brittle behavior. To overcome this disadvantage, the addition of appropriate fillers to brittle matrices is recommended [14, 22, 32, 35]. For this objective, including fiber reinforcement into the geopolymer matrix can increase the mechanical performance of the final composite [12]. The evaluation of the mechanical properties of reinforced geopolymer composites indicated that adding fibers helps to limit the propagation of micro- and macro-cracks across the material, as well as control shrinkage stresses, especially when exposed to high temperatures [9, 20, 23]. Various natural fibers, including cotton [33], sisal [36], coconut [30], bamboo [25, 30], flax [8], hemp [37], and banana [21], have been successfully incorporated into geopolymer matrices. The optimal fiber content for enhanced mechanical properties of reinforced geopolymer is typically around 0.5–1.0 wt% [38]. However, increasing fiber content may lead to decreased workability, density, and compressive strength. Before the synthesis of fiber-reinforced geopolymer composites, the selection of natural fibers that have desirable features for the specific target should be performed. Then, preparation of the natural fiber with either physical/mechanical or chemical methods is required based on the nature of the fiber source [39]. The performance of prepared fiber is also characterized by mechanical, chemical, thermal, and other properties.

Once the selection and preparation of natural fiber are completed, the next step will be the preparation of the geopolymer matrix. The aluminosilicate source can be from natural sources (clays), industrial wastes, and ashes mixed with an activator in a pre-set proportion. Then the prepared natural fiber is introduced to the geopolymer matrix at the appropriate condition, usually at room temperature [40, 41]. This low-temperature processing of geopolymers allows for the use of organic fibers without thermal degradation, although the highly alkaline matrix may potentially degrade the fibers [13]. After the preparation of reinforced geopolymer, it should be aged at a given temperature and humidity for some time, such as 7 and 28 days, at which the interaction of components within the matrix improves [40, 42]. The quality of the final product is influenced by the degree of mixing, the amount of fiber, the concentration of the activator, the type of aluminosilicate source used, the curing condition, etc. Alomayri and Low [39] synthesized a cotton fiber-reinforced geopolymer composite with an alkaline solution to fly ash ratio of 0.35. Four geopolymer composites with varying cotton fiber content were developed and mixed with dry fly ash until homogeneity was achieved, then an alkaline solution was added. The mixture was cast into 25 silicon molds, cured at 105°C for 3 hours, and then air-dried for 28 days. Another study [38] investigated the addition of sisal, coconut, and glass fiber to geopolymer mortars, focusing on their volume fractions and workability, as high fiber volumes can hinder casting and compacting. The fly ash, sodium hydroxide, river sand, sodium silicate, and fiber were well mixed. The geopolymer mortar was cast into molds, vibrated for compaction, and stored at room temperature. Then it was cured in the oven at 60°C for 48 hours and then de-molded (**Figure 1**).

In the synthesis of natural fiber-reinforced geopolymer, the fiber-matrix bonding should be considered as a major controlling factor. Ensuring good adhesion between fibers and the geopolymer matrix is crucial for the mechanical performance of the composite. Surface treatments or coupling agents may be used to enhance bonding. The other basic issue is sustainability. Both natural fibers and geopolymers are chosen for their environmental benefits, so it's important to ensure that the entire process aligns with sustainability goals.

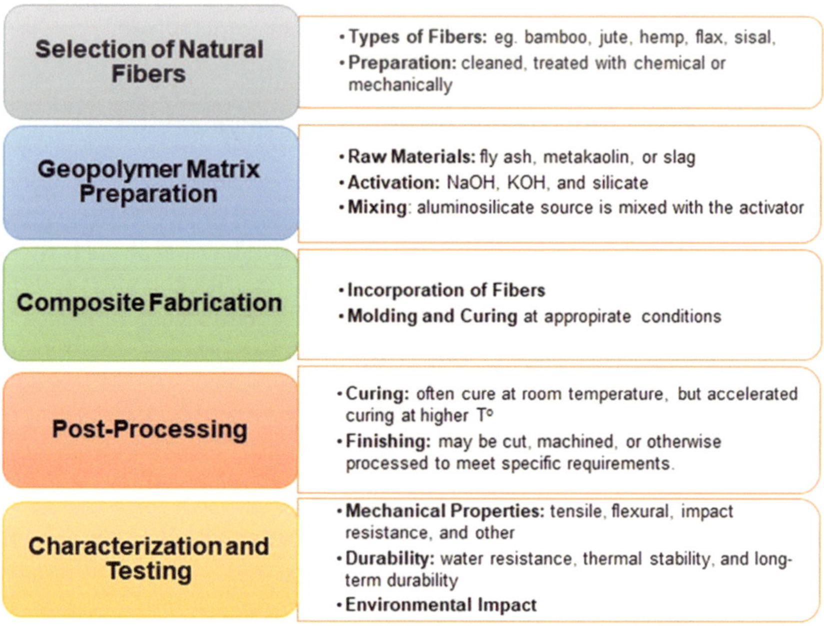

Figure 1.
Synthesis of geopolymer reinforced with natural fiber.

4. Effect of natural fiber reinforcement on the performances of geopolymer composites

Like cementitious matrices (such as cementitious matrices of pastes, mortars, and concrete), geopolymers show high brittle behavior and have low tensile and flexural strengths [20]. Furthermore, geopolymer materials demonstrate a significant level of shrinkage, although research shows that it is less pronounced than that observed in Portland cement-based materials [42]. The common approach to overcome such drawbacks is the addition of suitable fillers to the matrices. For this purpose, fiber reinforcement of geopolymer matrices can improve the performances of the resulting composites. Fiber reinforcement of geopolymer composites helps to reduce and control the propagation of micro- and macro-cracks through the material and shrinkage stresses [43, 44]. Both synthetic (e.g., polypropylene, polyvinyl alcohol, and carbon, basalt, steel, and glass) [17, 43–49] and natural (hemp, jute, kenaf, flax, false banana, and wood) [9, 50–58] fiber reinforcements have already been investigated. However, the use of natural fibers as reinforcement has grown fast as cost-effective, friendly, and renewable materials due to an increasing environmental concern from synthetic fibers. Natural fibers have been considered the materials of the future as reinforcement in geopolymer composites because of their excellent mechanical properties, wide availability, cost-effectiveness, biodegradability, non-toxicity, and low density.

In producing geopolymers, developing the optimum mix of the components, considering the alkali activator, the NaOH concentration, the cement/binder, and the aluminosilicate, is critical. Then, to produce the fiber-reinforced geopolymer, the fiber with proper length and weight percentage is introduced to the optimum mix design. The fiber is first dry-mixed with the binder and then with the alkaline activator solution to ensure the homogeneity of the composite. The thoroughly mixed geopolymer paste is poured into a mold with specified geometry and tapped to ensure optimal compaction and to expel trapped air [9].

4.1 Characterization of natural fiber-reinforced geopolymer binder

Composite materials are mainly made up of two basic elements, the matrix, or continuous phase, and a reinforcing material, or dispersed phase, or filler dispersed in the matrix, and the interface separates them [59, 60]. The interphase is the region of interaction between the reinforcement and the matrix with chemical and physical bonds [54]. The function of the matrix is to protect the reinforcement from a harsh or corrosive environment, keep the reinforcement in place, and transfer tension to it. The interface is a hypothetical surface without thickness joining the reinforcement and the matrix by a certain adhesion mechanism. Achieving adequate interface and interphase between matrix and reinforcement is the main challenge in processing composite materials.

Geopolymers are produced from different components with different sources for each component, i.e., there are many types of aluminosilicate sources and activated agents or binders that can be used as raw materials to produce geopolymers. The chemical compositions of the raw materials and other synthesis factors, including the optimum mix design of the starting materials and the processing conditions (e.g., curing time), as well as the addition of fillers/reinforcements, can influence the final properties of geopolymers. The common properties of the geopolymer pastes, mechanical, thermal, and water behavior, are discussed below.

4.1.1 Mechanical performance

Geopolymers as future prospective construction materials are mainly characterized by their mechanical properties (e.g., compressive strength, splitting tensile strength, flexural strength, impact resistance) as their most important performance indicators. Studies demonstrated that geopolymer concretes (e.g., made of metakaolin) exhibit higher mechanical strengths than ordinary Portland cement concretes [61, 62]. Geopolymers exhibit different mechanical properties depending on the source of the binder material. For instance, metakaolin-based geopolymers show good mechanical properties at room temperature, which are likely to decrease after exposure to high temperatures, in contrast to FA-based geopolymers, which have a much lower strength at an early age but better mechanical properties at elevated temperatures [6, 63, 64]. This could be due to the difference in the degree of polymerization with respect to the standing time, which for fly ash-based geopolymers increases with time when coupled with optimum heat curing. On the other hand, it has been reported that the Si/Al ratio in the source aluminosilicate material affects the mechanical performance of geopolymers [65–69]. An optimum Si/Al ratio (~2) promotes the geopolymerization reaction to generate more gel phase and contributes to the formation of more Si-O-Si and Si-O-Al bonds. A decrease in mechanical performance of geopolymers at a higher Si/Al ratio may be related to the lack of Al, which leads to a decrease in the structural reactivity of the geopolymer and adversely affects the structural reorientation. The Si/Al ratio from 1.5 to 2 produced a 3D network, while increasing the Si/Al ratio to more than three times formed the low-compressive-strength two-dimensional cross-linked polysialate [22, 65]. Furthermore, Wan et al. [66] reported that the high Si/Al ratio results in a large number of unreacted or partially reacted particles in the mixture, which causes an agglomeration phenomenon that affects the strength of geopolymers. Some research demonstrated that mixing such precursors with different temporal responses may result in geopolymers with optimal strength, either at room temperature or after exposure to elevated temperatures [62, 70].

Moreover, previous studies investigated that type of activator affects the mechanical strength of geopolymers [71]. In general, acidic activating agents exhibit higher compressive strength than alkali-based geopolymers [72]. For the alkali activation, Na activators promote higher dissolution of the aluminosilicate precursor to liberate amorphous Si and Al than K and result in a faster reaction rate. Hence, Na-based geopolymers provide higher strength than K-based geopolymers [71]. Besides, the molar concentration of the activating agent (acidic or alkaline) also affects the performance of geopolymer concretes significantly; mostly, maximum strength is obtained at 10–16 M [72]. A reduction in strength at very high concentrations (≥16 M) is observed due to the excess of hydroxide groups, which caused the precipitation of aluminosilicate gel before the initiation of the geopolymerization process. Generally, a strongly alkaline medium is required to dissolve and hydrolyze Si and Al amorphous phases.

On the other hand, the mechanical properties of geopolymers are strongly influenced by the starting raw mineral materials and curing process, which control the microstructural variations such as density, porosity, pore size, pore distribution, and micro-crack distribution [73, 74]. Microcracks readily grow around large pores. High curing temperatures may improve the mechanical strength of geopolymers synthesized from certain binders. However, the recent trend on alleviating these challenges involves natural fiber reinforcement of geopolymer pastes, which limits the crack initiation and propagation rate [74]. The effectiveness of natural fibers on reduction of defects significantly depends on the physicochemical characteristics of the fiber, such as the amount added, stiffness, length, orientation, and fiber-binder interfacial bonding. Energy absorption by the fiber is the main mechanism of limiting flaws in geopolymer pastes, i.e., when a propagating crack faces a fiber, a higher energy of fracture is required to pull-out or rupture the fiber and to advance the crack propagation [9]. Fiber reinforcement of fragility matrices as a mechanical characteristic increases the toughness, ductility of the material, compressive strength, and bending resistance since the fibers control the propagation of cracks in the material under different loadings and also in cases where retraction occurs [75].

The performances of natural fiber-reinforced geopolymers strongly depend on the interfacial bonding between the fiber and the paste [76], the intensity of which is determined by the physical and chemical characteristics of the fiber used. The interphase region can be a diffusion zone, a nucleation zone, a chemical reaction zone, or any combination of these parameters [77]. The interphase region is developed by the interactions of the matrix with the surface of the reinforcement [77, 78]. Both the interface and the interphase, mainly the adhesion, contribute to the mechanical properties of the composite. The adhesion mechanism may be interaction of the fiber surface with the matrix in electrostatic attraction (due to surface charge differences between the matrix and the fiber), formation of chemical bonds (formed between chemical groups on the reinforcement surface and compatible groups on the matrix), or inter-diffusion of materials [57, 78–81]. These adhesion mechanisms between the matrix and the reinforcement significantly contribute to the mechanical properties through the transfer of loads between the matrix and the reinforcement.

Raw natural fibers usually exhibit weak interfacial bonding with geopolymer pastes due to their smooth or corrugated surfaces. Employing various pretreatment techniques (chemical or thermochemical) of fibers removes undesirable substances covering the external surface (e.g., waxes, oils, lignin, etc.) and significantly improves fiber-geopolymer bonding [9, 51]. Alkaline treatment with appropriate concentration is one of the most effective and widely used treatment methods [9, 51, 52, 54, 55, 82]. Ye et al. [58] reported that the interfacial bonding properties of wood-geopolymer composites were improved by increasing the roughness of the wood surface through alkaline treatment of sisal fiber. Addis et al. [9] investigated the FTIR spectra of untreated and alkali-treated (NaOH) false banana fibers and correlated the effect of treatment on interfacial bonding with the absorption band characteristics. Significant increase in hydroxyl (OH) functional groups on the surfaces of the fibers for the alkali-treated boosted the fiber's reactivity and promoted better adhesion with the matrix material, resulting in improved mechanical properties of fiber-reinforced geopolymer composites. Furtos et al. [83] reported that the presence of hydroxyl groups in wood fibers can form hydroxide bridges, and hydrogen bonding may play an important role in the adhesion between wood fibers and the geopolymer matrix.

Moreover, research shows that the amount of fiber added significantly influences the mechanical strength of geopolymer composites. Poor fiber distribution limits the reinforcement role of the fiber at low fiber loading. At higher fiber content, the fibers will be uniformly dispersed in the matrix and enhance the mechanical strength of geopolymer composites [84]. At significantly higher fiber loading, void formation and poor fiber dispersion reduce the reinforcement efficacy of the fiber. Most commonly, increasing natural fiber content up to 1 wt% leads to improved mechanical strength of plain geopolymer composites regardless of fiber type [55, 56, 85]. However, higher proportions of treated natural fiber can be added as a reinforcement (up to 3 wt%)

compared to raw fibers and maximize the mechanical strength of the composites [50, 86]. This may be related to the improved dispersion and interfacial adhesion of the fiber with the geopolymer paste. Furthermore, fiber length is another important factor that affects the mechanical performance of fiber-reinforced geopolymer composites. It has been observed that an optimum fiber length maximizes the performance of the composites. Longer fibers reduce the mechanical strength of geopolymer composites due to the entanglement of fibers and subsequent formation of agglomerates, which occupy a particular space in the composite and cause an increase in voids, resulting in weak fiber-geopolymer interfacial bonding [53, 84].

In summary, the enhancement in mechanical performance of natural fiber-reinforced geopolymer composites comes mainly from the extensive energy-absorbing characteristics through fiber breakage, fiber-matrix interface delamination, fiber pull-out, and fiber bridging, which slow crack propagation and thus increase fracture energy. Insufficient interfacial bonding length and low fiber stiffness may initiate a fiber rupture and the crack propagates in the geopolymer binder. Furthermore, fiber orientation and the direction of applied forces significantly control the fracture energy. Fiber alignment in the direction of efficiently bridge cracks results in better mechanical performance [87]. If the crack is propagated in the plane perpendicular to the fiber, this latter encounters the fracture energy that is released to form cracks by the bridging mechanism. Otherwise, the crack continues to propagate by passing the fiber, as illustrated in **Figure 2**.

4.1.2 Thermal performance

Geopolymeric materials are generally characterized by their low thermal conductivity compared to conventional building materials [65, 89–92]. Most importantly, the thermal conductivity of geopolymers is largely sensitive to the type of binder (i.e., mainly the Si/Al ratio) [90]. An increase in the Si/Al ratio decreases the amount and size of pores, increasing the thermal conductivity of the geopolymeric

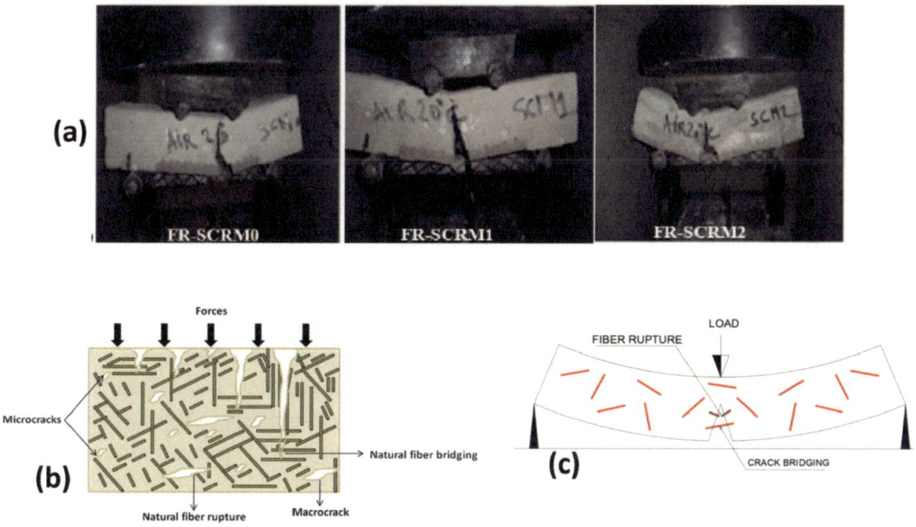

Figure 2.
(a) Bridging mechanism in natural fiber-geopolymer composite; (b) Fiber organization scheme in a crack; (c) Breakage of specimens with fiber [20, 73, 88].

materials [65, 91]. Moreover, the alkali activator type affects the thermal conductivity of geopolymers. The transfer of alkali ions ($Na+$, $K+$) proportion into the networks to balance the negative charge in aluminum-centered sites reduces the network's porosity [73]. Duxson et al. [92] found that the thermal conductivity of Na-geopolymers is greater than K-geopolymers due to the lower specific heat of potassium compared to sodium.

Very limited research studied the effect of incorporation of natural fibers in geopolymer composites on their thermal properties. Similar to the mechanical properties, the effect of the fiber on the thermal property strongly depends on the treatment of the fiber. Most of the findings show that untreated fibers reduce the thermal conductivity of the geopolymer composites, mainly attributed to the low chemical interaction between untreated fibers and the geopolymer matrix, resulting in increased air voids and, consequently, a lower thermal conductivity [86]. Besides, an increase in fiber content and length results in a reduction in thermal conductivity, which may be correlated to the low thermal conductivity of the fibers, attributed to the packing of the fibers in the geopolymer matrix and the structure of the fibers [4].

4.1.3 Density and porosity

Natural fiber reinforcement of geopolymer binders profoundly affects the density and the porosity of the composites. Most studies indicate that an increase in fiber loading increases the porosity of geopolymer composites and decreases the density [39]. The decrease in density is attributed to the weak fiber-matrix bonding, which generates void areas surrounding natural fibers, and to the agglomeration of the fibers during mixing, thus trapping water-filled spaces that turn into voids. Treatment of the fiber also influences the effect of the fiber on the density of the composites; untreated natural fibers cause a slight decrease on the density of geopolymer composites compared to treated fibers, which can show a moderate or relatively lesser effect on the density of geopolymer composites [93]. The higher contents of hemicellulose and lignin in untreated natural fibers are responsible for this effect, which influences negatively the fiber-matrix interface and creates more pores, resulting in low density [93].

4.1.4 Hydric performance

Hydric properties (i.e., water absorption, water permeability) have significant impacts on the quality and durability of geopolymer composites. High water absorption and permeability lead to a variety of degradation mechanisms in geopolymer composites, including freeze-thaw damage, alkali-silica reaction, and corrosion of steel reinforcement. The factors that critically determine the hydric properties of geopolymers are the particle size distribution of raw materials, curing conditions, the Si/Al ratio, and active agent type [94–97]. Fine particles fill vacant spaces and produce composites with higher density and consequently more resistance to water permeability [94]. High-temperature curing increases the percentage of water absorption due to the dehydration and growth of larger pores in the composite [95]. Research shows that water absorption of geopolymers increases at low Si/Al ratio (1.6–3) and decreases at high Si/Al ratio (≥ 3) [96, 98]. The decrease in water absorption at high Si/Al ratio is attributed to the higher distribution of small pores that are isolated, and the impact of their presence is not significant. When considering the effect of activator type, studies revealed that Na-based geopolymers exhibit better resistance to water absorption than K-based geopolymers, which can be attributed to the refinement of the surface pores [97].

For natural fiber-reinforced geopolymer composites, there is limited literature on the influence of the natural fiber on the hydric properties of the composites. The hygroscopic behavior of natural fibers is an inherent characteristic that can influence their applications as reinforcements in geopolymer composites. The fiber orientation, fiber nature (permeable or not), fiber content, exposed surface area, diffusion, and type of surface protection determine the hydric properties of the composites [12, 50, 99–102]. The pores in the geopolymeric composites are responsible for the permeability of moisture and gases from the environment into the matrix, reaching the natural fiber reinforcement; this modifies the matrix structure, the natural fiber reinforcement, and consequently the durability of the fiber and the composite as a whole [103–106]. Most commonly, the available literature shows that the water absorption of the composite does not exhibit significant change due to the fiber addition [12]. Some research outputs reported that the water absorption decreases with increasing fiber length and fiber loading due to the high solid content of the composite and more dense packing of short fibers [50, 99]. In summary, further studies are needed to elucidate the effect of natural fibers (fiber stiffness, fiber orientation, fiber size, and fiber content) on the long-term durability of a natural fiber-reinforced geopolymer composite.

5. Natural fibers vs. synthetic fibers for geopolymer reinforcement

Recently, fiber-reinforced geopolymer composites are playing an important role in recent times in the automobile, aerospace, and naval architecture industries as an advanced technological solution [85]. Synthetic fibers (such as steel, glass, and carbon fibers) have been the most common fibers in such geopolymer reinforcements [107–109]. Besides, synthetic polymeric fibers (such as polypropylene and polyvinyl alcohol fibers) have been studied for reinforcement applications [110, 111]. Ilamvazhuthi and Gopalakrishna [112] evaluated the compressive strength of a polypropylene fiber-reinforced geopolymer concrete as a function of wet mixing time and compared it with traditional cement concrete with the fiberized geopolymer (**Figure 3**).

Natural fiber reinforcement of geopolymer composites is mainly motivated by the effect of synthetic fibers on the environment, aiming to produce sustainable building materials due to the renewable, biodegradable, and non-toxic nature of natural fibers and is a recent phenomenon. Furthermore, natural fibers are advantageous for their low density accompanied by comparable mechanical strength and for their economic gains given that the production of synthetic fibers is an energy-consuming process, unlike the planting and harvesting of natural fibers [55, 85, 113–115].

Most importantly, the performance of a geopolymeric material reinforced by both natural and synthetic fibers strongly depends on the fiber properties, the fiber content, the geopolymeric precursor, the curing conditions, and the setting time [42]. Even more important is the interfacial bonding between the fiber surface and the geopolymer matrix, which controls the mechanism of the load transfer from the matrix to the fiber, significantly determining the performance of the geopolymeric

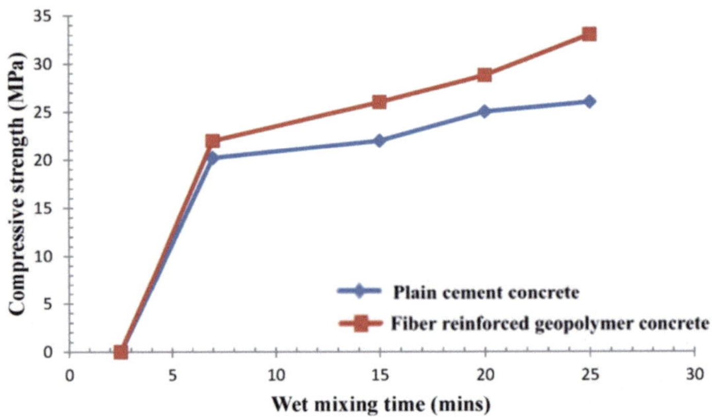

Figure 3.
Effect of mixing time on the compressive strength of traditional and geopolymer concrete [112].

composite [110, 116, 117]. The surface nature of fibers (smooth or rough) plays a considerable role in determining the intensity of the interfacial bonding between the fiber and the matrix, hence defining the performance of the composite. Ranjbar N, Zhang [74] used steel microfibers and polypropylene fibers to analyze the interfacial bond geopolymeric pastes and found that the polypropylene fiber exhibits weak adhesion owing to its smoother and more regular surface, while steel microfiber showed strong adhesion with the matrix due to a much rougher surface. This was further proved by investigating the performances of the composite samples by comparing the results of the flexural strength tests; composites reinforced with steel microfibers had higher strengths than the polypropylene fiber-reinforced composites. These differences in performance are further aggravated by the differences in hydroscopic properties of the fibers. The surface of the polypropylene fiber is hydrophobic while that of the steel microfiber is hydrophilic, leading to a weak fiber-matrix interface for the former and further deterring the performance of the composites.

On the other hand, for natural fiber-reinforced geopolymer composites, although the hygroscopic properties and the surface morphology of the fiber surface are still important factors, other characteristics (mainly the chemical composition of the fiber, i.e., the lignin, hemicellulose, and cellulose contents) play more important roles in the fiber-matrix adhesion [118]. Non-cellulosic organic components (i.e., hemicellulose and lignin) are sensitive to highly alkaline environments, and they tend to degrade under these conditions and delay the hardening of the paste due to the generation of carboxylic acid as a product inhibiting the geopolymerization reactions, which not only decreases the fiber's compatibility with the geopolymer matrix but also causes a temporal reduction in the composite's performance [55, 93, 119]. Furthermore, lignin and hemicellulose cause an increase in porosity, which further reduces the performance of the composites. The cellulose component of the fiber is not sensitive to an alkaline medium and does not undergo significant degradation in it; it also has a tendency to create improved adhesion between the fiber and the geopolymer matrix, which results in reduced porosity and consequently improved performance of the composites [93]. Similar to synthetic fibers, fiber-matrix adhesion in natural fiber-reinforced geopolymers is affected by the hygroscopic properties of the fibers [120]. Natural fibers are hydrophilic by nature due to the presence of hydroxyls and other polar groups on the surfaces, giving them the characteristics of high moisture absorption and little dimensional stability (swell in contact with water). This behavior leads not only to the weakening of the fiber-matrix adhesion but also to the formation of cracks, resulting from the drying of the composite, and a reduction in mechanical performance [120, 121]. Considering all the factors analyzed above in relation to the effectiveness of the fiber-matrix adhesion, most of the studies recommend treatment of natural fibers to modify the chemical composition and the surface chemistry so that the fiber-matrix interface is optimized.

6. Conclusions perspectives

Geopolymers, offering several advantages such as higher strength, a lower carbon footprint, a lighter weight, and increased durability, have emerged as promising alternatives to traditional cement-based materials for building and construction applications. The important factors determining the synthesis of geopolymers include the types of the aluminosilicate sources, synthesis processes, curing conditions, setting time, and the reinforcement mechanism, which considerably determine the microscopic and macroscopic properties of the composites. In addition to producing sustainable building and construction materials in order to meet the state-of-the-art in terms of environmental protection, natural fiber reinforcement of geopolymers provides meaningful gains in improving their mechanical properties.

Research findings in the last few decades demonstrate that the mechanical, thermal, and hydric properties of the natural fiber-reinforced geopolymer composites are the most important performance indicators and are strongly dependent on the type of reinforcement fiber, content, size, its mechanical properties, and treatment conditions, which significantly determine fiber-matrix adhesion intensity and consequently the performance of the composites. Many of the findings reported that flaws or defects are common when raw natural fibers are used for geopolymer reinforcement, mainly coming from the poor fiber-matrix adhesion and agglomeration of the fibers; this results in composites with increased porosity and cracks deterring their performances. Hence, implementing an appropriate treatment of the natural fibers is critical to improve the fiber-matrix interfacial adhesion, consequently to enhance the mechanical, thermal, and hydric properties of the natural fiber-reinforced geopolymer composites. In this regard, further studies are required to illustrate the economic feasibility of applying natural fibers in geopolymers as reinforcement since the fiber treatment incurs additional manufacturing costs of the composites. Furthermore, the long-term durability, the optimum composition, the optimum processing conditions, the appropriate curing conditions, and the tangible environmental gains of natural fiber-reinforced geopolymer composites require further studies. Provided that the above-mentioned concerns are solved, applications of natural fiber-reinforced geopolymer composites present implausible promises as sustainable building materials. Hence, it can be concluded that, with continued research and development, natural fiber-reinforced geopolymer composites can offer a sustainable and viable alternative to traditional building materials in the near future.

Author details

Zenamarkos Bantie* and Desalegn Abera
Chemical Engineering Program, Faculty of Chemical and Food Engineering,
Bahir Dar Institute of Technology-Bahir Dar University, Bahir Dar, Ethiopia

*Address all correspondence to: zenamarkosbs@gmail.com
and zenamarkos.bantie@bdu.edu.et

References

[1] Majidi B. Geopolymer technology, from fundamentals to advanced applications: A review. Materials and Technologies. 2009;**24**:79-87. DOI: 10.1179/175355509X449355

[2] Olayiwola HO, Amiandamhen SO, Meincken M, Tyhoda L. Investigating the suitability of fly ash/metakaolin-based geopolymers reinforced with south African alien invasive wood and sugarcane bagasse residues for use in outdoor conditions. European Journal of Wood and Wood Products. 2021;**79**:611-627. DOI: 10.1007/s00107-020-01636-4

[3] Adewuyi YG. Recent advances in fly-ash-based geopolymers: Potential on the utilization for sustainable environmental remediation. ACS Omega. 2021;**6**:15532-15542

[4] Walbrück K, Drewler L, Witzleben S, Stephan D. Factors influencing thermal conductivity and compressive strength of natural fiber-reinforced geopolymer foams. Open Ceramics. 2021;**5**:100065. DOI: 10.1016/j.oceram.2021.100065

[5] Dai X, Ren Q, Aydın S, Yardımcı MY, Lesage K, De Schutter G. Enhancing thixotropy and structural build-up of alkali-activated slag/fly ash pastes with nano clay. Materials and Structures. 2021;**54**:1-17. DOI: 10.1617/s11527-021-01760-4

[6] Noushini A, Hastings M, Castel A, Aslani F. Mechanical and flexural performance of synthetic fiber reinforced geopolymer concrete. Construction and Building Materials. 2018;**186**:454-475. DOI: 10.1016/j.conbuildmat.2018.07.110

[7] Mohseni E. Assessment of Na2SiO3 to NaOH ratio impact on the performance of polypropylene fiber-reinforced geopolymer composites. Construction and Building Materials. 2018;**186**:904-911. DOI: 10.1016/j.conbuildmat.2018.08.032

[8] Silva G, Kim S, Bertolotti B, Nakamatsu J, Aguilar R. Optimization of a reinforced geopolymer composite using natural fibers and construction wastes. Construction and Building Materials. 2020;**258**:119697. DOI: 10.1016/j.conbuildmat.2020.119697

[9] Addis LB, Sendekie ZB, Habtu NG, Schubert DW, Roether JA, Boccaccini AR. False banana fiber reinforced geopolymer composite – A novel sustainable material. Ceramics International. 2024;**50**:24442-24450. DOI: 10.1016/j.ceramint.2024.04.177

[10] Camargo MM, Taye EA, Roether JA, Redda DT, Boccaccini AR. A review on natural fiber-reinforced geopolymer and cement-based composites. Materials (Basel). 2020;**13**:1-29

[11] Li Y, Maimait A, Cheng J, Duan Y, Chen Y, Dong H. Properties of Fiber-reinforced Geopolymer mortar using coal gangue and Aeolian sand. Materials (Basel). 2024;**17**:3225. DOI: 10.3390/ma17133225

[12] Saccani A, Molari L, Totaro G, Manzi S. Geopolymers reinforced with natural fibers: A comparison among different sources. Applied Sciences. 2021;**11**:11026. DOI: 10.3390/app112211026

[13] Addis LB, Sendekie ZB, Satheesh N. Degradation kinetics and durability enhancement strategies of cellulosic fiber-reinforced geopolymers and cement composites. Advances in Materials Science and Engineering. 2022;**2022**:1981755

[14] Zulfiati R, Saloma IY. Mechanical properties of fly ash-based geopolymer with natural fiber. Journal of Physics: Conference Series. 2019;**1198**:082021

[15] Qu Y, Li W, Wang C, Cao H, Zhou R. Study on the reinforced properties of geopolymer fibers with a sustainable development role. Sustainability. 2024;**16**:6255

[16] Dong X, Zhou H, Jiang W, Shi K. Flexural capacity of aligned steel fiber-reinforced geopolymer composite beams. AIP Advances. 2024;**14**:1-8. DOI: 10.1063/5.0214832

[17] Baziak A, Pławecka K, Hager I, Castel A, Korniejenko K. Development and characterization of lightweight geopolymer composite reinforced with hybrid carbon and steel. Materials (Basel). 2021;**14**:5741. DOI: 10.3390/ma14195741

[18] Ascens G, Bernardo E. An investigation on the synthesis of alkali activated materials from thermally modified clays guilherme. Applied Sciences. 2022;**12**:9085. DOI: 10.3390/app12189085 Academic

[19] Assaedi H, Shaikh FUA, Low IM. Effect of nano-clay on mechanical and thermal properties of geopolymer. Journal of Asian Ceramic Societies. 2015;**4**:1-10. DOI: 10.1016/j.jascer.2015.10.004

[20] de Azevedo ARG, Cruz ASA, Marvila MT, de Oliveira LB, Monteiro SN, Vieira CMF, et al. Natural fibers as an alternative to synthetic fibers in reinforcement of geopolymer matrices: A comparative review. Polymers (Basel). 2021;**13**:2493

[21] Rahman MM, Payel MTH, Asaduzzaman M, Hossain S, Ali M. Physical properties of isolated cellulose fiber from jute and banana fiber through Kraft pulping: Potential applications in packaging and regenerated fibers. SPE Polymers. 2024;**1-9**. DOI: 10.1002/pls2.10155

[22] Kaushik Y, Sooriyaperakasam N, Rathee U, Naik N. A mini review of natural cellulosic fibers: Extraction, treatment and characterization methods. Journal of Computers, Mechanical and Management. 2023;**2**:1-16. DOI: 10.57159/gadl.jcmm.2.2.23057

[23] Bravo-Moncayo L, Argotti-Gómez M, Jara O, Puyana-Romero V, Ciaburro G, Guerrero VH. Thermo-acoustic properties of four natural fibers, *Musa textilis*, *Furcraea andina*, *Cocos nucifera*, and *Schoenoplectus californicus*, for building applications. Buildings. 2024;**14**:2265. DOI: 10.3390/buildings14082265

[24] Meziane H, Laita M, Azzaoui K, Boulouiz A, Neffa M, Sabbahi R, et al. Nanocellulose fibers: A review of preparation methods, characterization techniques, and reinforcement applications. Moroccan Journal of Chemistry. 2024;**12**:305-343. DOI: 10.48317/IMIST.PRSM/morjchem-v12i1.44573

[25] Geremew A, De Winne P, Demissie TA, De Backer H. Surface modification of bamboo fibers through alkaline treatment: Morphological and physical characterization for composite reinforcement. Journal of Engineered Fibers and Fabrics. 2024;**19**:1-13. DOI: 10.1177/15589250241248764

[26] Azalam Y, Alioui A, Al Armouzi N, Benfars M, Mabrouki M, Bendada EM. Physical and mechanical properties of adobe bricks reinforced by natural additives, a case study of alfalfa fibers. EUREKA Physics and Engineering. 2024, 2024:144-159. DOI: 10.21303/2461-4262.2024.003426

[27] Blal A, Brouillette F, Loranger É, Lebrun G. Click chemistry modifications for the selective crosslinking of wood pulp fibers - effect on the physical and mechanical properties of paper. RSC Advances. 2024;**14**:9656-9667. DOI: 10.1039/d3ra08590b

[28] Moussaoui N, Benhamadouche L, Seki Y, Amroune S, Dufresne A, Jawaid M, et al. The impact of physicochemical treatments on the characteristics of *Ampelodesmos mauritanicus* plant fibers. Cellulose. 2023;**30**:7479-7495. DOI: 10.1007/s10570-023-05377-4

[29] Razali N, Salit MS, Jawaid M, Ishak MR, Lazim Y. A study on chemical composition, physical, tensile, morphological, and thermal properties of roselle fiber: Effect of fiber maturity. BioResources. 2015;**10**:1803-1823. DOI: 10.15376/biores.10.1.1803-1824

[30] dos Santos AJG, Ribeiro MM, de Alessandro Corrêa C, da Jean Rodrigues S, Silva DS, Raí FP. Morphological, chemical and mechanical properties of hybrid polyester composites reinforced with bamboo fibers and kaolin waste. Journal of Materials Research and Technology. 2024;**30**:1-15. DOI: 10.1016/j.jmrt.2024.03.003

[31] Varma VK, Sarangi SK. Characterization of *Bauhinia variegata* natural fiber treated with bleaching, benzoylation, and alkalization. Engineering Research Express. 2024;**6**:025528. DOI: 10.1088/2631-8695/ad3acc

[32] Fernández MV, Rivera JLV, Rodríguez FP, Losada HF, Abreu MEF, Diaz FRV, et al. Mechanical characterization of a polymer/natural fibers/bentonite composite material with implementation of a continuous damage model. Applied Sciences. 2023;**13**:2677. DOI: 10.3390/app13042677

[33] Nam S, Condon BD, Delhom CD, Fontenot KR. Silver-cotton nanocomposites: Nano-design of microfibrillar structure causes morphological changes and increased tenacity. Scientific Reports. 2016;**6**:1-10. DOI: 10.1038/srep37320

[34] Guler GS, Sumnu G, Yazicioglu N. Caffeic acid encapsulated centrifugally spun antioxidant fibers: Characterization and incorporation in cakes. Food and Bioprocess Technology. 2024:1-19. DOI: 10.1007/s11947-024-03455-4

[35] Rinovian A, Amin M, Prasetyo E, Erlangga BD, Nulhakim L, Salahudin W. The effect of adding natural fibers on the geopolymer characterization and mechanical performance. AIP Conference Proceedings. 2024;**3003**:1-10. DOI: 10.1063/5.0186500

[36] Leite FL, Herrmann PSP, Da Róz AL, Ferreira FC, Curvelo AAS, Mattoso LHC. Investigation of sisal fibers by atomic force microscopy: Morphological and adhesive characteristics. Journal of Nanoscience and Nanotechnology. 2006;**6**:2354-2361. DOI: 10.1166/jnn.2006.522

[37] Sambucci M, Sibai A, Valente M. Recent advances in geopolymer technology. A potential eco-friendly solution in the construction materials industry: A review. Journal Of Composites Science. 2021;**5**:109

[38] Wongsa A, Kunthawatwong R, Naenudon S, Sata V, Chindaprasirt P. Natural fiber reinforced high calcium fly ash geopolymer mortar. Construction and Building Materials. 2020;**241**:118143. DOI: 10.1016/j.conbuildmat.2020.118143

[39] Alomayri T, Low IM. Synthesis and characterization of mechanical properties in cotton fiber-reinforced geopolymer composites. Journal of

Asian Ceramic Societies. 2013;**1**:30-34. DOI: 10.1016/j.jascer.2013.01.002

[40] Kani EN, Allahverdi A. Effects of curing time and temperature on strength development of inorganic polymeric binder based on natural pozzolan. Journal of Materials Science. 2009;**44**:3088-3097. DOI: 10.1007/s10853-009-3411-1

[41] Ramakrishnan S, Pasupathy K, Sanjayan J. Synthesis and properties of thermally enhanced aerated geopolymer concrete using form-stable phase change composite. Journal of Building Engineering. 2021;**40**:102756. DOI: 10.1016/j.jobe.2021.102756

[42] Amjad SU, Budihardjo MA, Priyambada IB, Asghar R. Synthesis of geopolymer concrete using flyash. IOP Conference Series: Earth and Environmental Science. 2022;**1098**:012072. DOI: 10.1088/1755-1315/1098/1/012072

[43] Aisheh YIA, Atrushi DS, Akeed MH, Qaidi S, Tayeh BA. Influence of polypropylene and steel fibers on the mechanical properties of ultra-highperformance fiber-reinforced geopolymer concrete. Case Studies in Construction Materials. 2022;**17**:e01234

[44] Aisheh YIA, Atrushi DS, Akeed MH, Qaidi S, Tayeh BA. Influence of steel fibers and microsilica on the mechanical properties of ultra-high-performance geopolymer concrete (UHP-GPC). Case Studies in Construction Materials. 2022;**17**:e01245

[45] Ganesh AC, Muthukannan M. Development of high performance sustainable optimized fiber reinforced geopolymer concrete and prediction of compressive strength. Journal of Cleaner Production. 2021;**282**:124543. DOI: 10.1016/j.jclepro.2020.124543

[46] Mastali M, Kinnunen P, Isomoisio H, Karhu M, Illikainen M. Mechanical and acoustic properties of fiber-reinforced alkali-activated slag foam concretes containing lightweight structural aggregates. Construction and Building Materials. 2018;**187**:371-381. DOI: 10.1016/j.conbuildmat.2018.07.228

[47] Krzywon R, Dawczynski S. Strength parameters of foamed Geopolymer reinforced with GFRP mesh. Materials. 2021;**14**:689. DOI: 10.3390/ma14030689

[48] Abbass M, Singh G. Impact strength of rice husk ash and basalt fibre based sustainable geopolymer concrete in rigid pavements. Materials Today Proceedings. 2022;**61**:250-257. DOI: 10.1016/j.matpr.2021.09.053

[49] Orouji M, Najaf E. Effect of GFRP rebars and polypropylene fibers on flexural strength in high-performance concrete beams with glass powder and microsilica. Case Studies in Construction Materials. 2023;**18**:e01769

[50] Bayraktar OY, Tobbala DE, Turkoglu M, Kaplan G, Tayeh BA. Hemp fiber reinforced one-part alkali-activated composites with expanded perlite: Mechanical properties, microstructure analysis and high-temperature resistance. Construction and Building Materials. 2023;**363**:129716. DOI: 10.1016/j.conbuildmat.2022.129716

[51] Suwan T, Maichin P, Fan M, Jitsangiam P, Tangchirapat W, Chindaprasirt P. Influence of alkalinity on self-treatment process of natural fiber and properties of its geopolymeric composites. Construction and Building Materials. 2022;**316**:125817. DOI: 10.1016/j.conbuildmat.2021.125817

[52] Silva G, Salirrosas J, Ruiz G, Kim S, Nakamatsu J, Aguilar R. Evaluation of

fire, high-temperature and water erosion resistance of fiber-reinforced lightweight pozzolana-based geopolymer mortars. IOP Conference Series: Materials Science and Engineering. 2019;**706**:012016. DOI: 10.1088/1757-899X/706/1/012016

[53] Abbas AGN, Aznieta AAFN, Khalina A, Azline MNN. Effects of fibers length on the properties of kenaf fibers- reinforced geopolymer concrete. AIP Conf. Proc. 2021;**2447**:030001. DOI: 10.1063/5.0072625

[54] Zhang N, Ye H, Pan D, Zhang Y. Effects of alkali-treated kenaf fiber on environmentally friendly geopolymer-kenaf composites: Black liquid as the regenerated activator of the geopolymer. Construction and Building Materials. 2021;**297**:123787. DOI: 10.1016/j.conbuildmat.2021.123787

[55] Lazorenko G, Kasprzhitskii A, Yavna V, Mischinenko V, Kukharskii A, Kruglikov A, et al. Effect of pre-treatment of flax tows on mechanical properties and microstructure of natural fiber reinforced geopolymer composites. Environmental Technology and Innovation. 2020;**20**:101105. DOI: 10.1016/j.eti.2020.101105

[56] Silva G, Kim S, Castañeda A, Donayre R, Nakamatsu J, Aguilar R, et al. A comparative study of linen (flax) fibers as reinforcement of fly ash and clay brick powder based geopolymers. IOP Conference Series: Materials Science and Engineering. 2018;**416**:012107. DOI: 10.1088/1757899X/416/1/012107

[57] Korniejenko K, Łach M, Hebdowska-Krupa M, Mikuła J. Impact of flax fiber reinforcement on mechanical properties of solid and foamed geopolymer concrete. Advances in Technology Innovation. 2020;**6**:11-20. DOI: 10.46604/aiti.2021.5294

[58] Ye H, Asante B, Schmidt G, Krause A, Zhang Y, Yu Z. Interfacial bonding properties of the eco-friendly geopolymer-wood composites: Influences of embedded wood depth, wood surface roughness, and moisture conditions. Journal of Materials Science. 2021;**56**:7420-7433. DOI: 10.1007/s10853-021-05775-8

[59] Loginov PA, Sidorenko DA, Orekhov AS, Levashov EA. A novel method for in situ TEM measurements of adhesion at the diamond–metal interface. Scientific Reports. 2021;**11**:1-10

[60] Fugolin APP, Costa AR, Correr-Sobrinho L, Crystal Chaw R, Lewis S, Ferracane JL, et al. Toughening and polymerization stress control in composites using thiourethane-treated fillers. Scientific Reports. 2021;**11**:1-12

[61] Almutairi AL, Tayeh BA, Adesina A, Isleem HF, Zeyad AM. Potential applications of geopolymer concrete in construction: A review. Case Studies in Construction Materials. 2021;**15**:e00733

[62] Raphaëlle P, Martin C. Formulation and performance of flash metakaolin geopolymer concretes. Construction and Building Materials. 2016;**120**:150-160. DOI: 10.1016/j.conbuildmat.2016.05.061

[63] Gambo S, Ibrahim K, Aliyu A, Ibrahim AG, Abdulsalam H. Performance of metakaolin based geopolymer concrete at elevated temperature, Niger. Journal of Technology. 2020;**39**:732-737. DOI: 10.4314/njt.v39i3.11

[64] Kohout J, Koutník P, Hajková P, Kohoutova E, Soukup A. Effect of K/Al molar ratio on the thermo-mechanical properties of metakaolinite-based geopolymer composites. Polymers. 2021;**13**:3754. DOI: 10.3390/polym13213754

[65] Ahmed MM, El-Naggar KAM, Tarek D, Ragab A, Sameh H, Zeyad AM, et al. Fabrication of thermal insulation geopolymer bricks using ferrosilicon slag and alumina waste. Case Studies in Construction Materials. 2021;**15**:e00737

[66] Wan Q, Rao F, Song S, García RE, Estrella RM, Patiño CL, et al. Geopolymerization reaction, microstructure and simulation of metakaolin-based geopolymers at extended Si/Al ratios. Cement and Concrete Composites. 2017;**79**:45-52. DOI: 10.1016/j.cemconcomp.2017.01.014

[67] Wang H, Wu H, Xing Z, Wang R, Dai S. The effect of various Si/Al, Na/Al molar ratios and free water on micromorphology and macro-strength of metakaolin-based geopolymer. Materials. 2021;**14**:3845. DOI: 10.3390/ma14143845

[68] Lahoti M, Narang P, Tan KH, Yang E-H. Mix design factors and strength prediction of metakaolin-based geopolymer. Ceramics International. 2017;**43**:11433-11441. DOI: 10.1016/j.ceramint.2017.06.006

[69] Tippayasam C, Boonsalee S, Sujjavanich S, Ponzoni C, Kamseu E, Chaysuwan D. Geopolymer development by powders of metakaolin and wastes in Thailand. Advances in Science and Technology. 2010;**69**:63-68. DOI: 10.4028/www.scientific.net/AST.69.63

[70] Tayeh BA, Zeyad AM, Agwa IS, Amin M. Effect of elevated temperatures on mechanical properties of lightweight geopolymer concrete. Case Studies in Construction Materials. 2021;**15**:e00673

[71] Omur T, Kabay N, Miyan N, Özkan H, Özkan Ç. The effect of alkaline activators and sand ratio on the physico-mechanical properties of blast furnace slag based mortars. Journal of Building Engineering. 2022;**58**:104998. DOI: 10.1016/j.jobe.2022.104998

[72] Huseien GF, Ismail M, Khalid NHA, Hussin MW, Mirza J. Compressive strength and microstructure of assorted wastes incorporated geopolymer mortars: Effect of solution molarity. Alexandria Engineering Journal. 2018;**57**:3375-3386. DOI: 10.1016/j.aej.2018.07.011

[73] Moujoud Z, Sair S, Ait Ousaleh H, Ayouch I, El Bouari A, Tanane O. Geopolymer composites reinforced with natural Fibers: A review of recent advances in processing and properties. Construction and Building Materials. 2023;**388**:131666. DOI: 10.1016/j.conbuildmat.2023.131666

[74] Ranjbar N, Zhang M. Fiber-reinforced geopolymer composites: A review. Cement and Concrete Composites. 2020;**107**:103498. DOI: 10.1016/j.cemconcomp.2019.103498

[75] Abu Alfoul B, Al Hattamleh O, Aldeeky H. Investigating the effects of disc shaped thumb tacks as a "fiber" reinforcement on various concrete properties. Case Studies in Construction Materials. 2019;**11**:e00257

[76] Todkar SS, Patil SA. Review on mechanical properties evaluation of pineapple leaf fibre (PALF) reinforced polymer composites. Composites. Part B, Engineering. 2019;**174**:106927

[77] Ahmad MR, Chen B, Haque MA, Saleem Kazmi SM, Munir MJ. Development of plant-concrete composites containing pretreated corn stalk bio-aggregates and different type of binders. Cement and Concrete Composites. 2021;**121**:104054

[78] Lyu B-C, Ding C, Guo L-P, Chen B, Wang A-G. Basic performances and potential research problems of strain

hardening geopolymer composites: A critical review. Construction and Building Materials. 2021;**287**:123030

[79] Gassan J. A study of fibre and interface parameters affecting the fatigue behaviour of natural fibre composites. Composites. Part A, Applied Science and Manufacturing. 2002;**33**:369-374

[80] El Alouani M, Saufi H, Moutaoukil G, Alehyen S, Nematollahi B, Belmaghraoui W, et al. Application of geopolymers for treatment of water contaminated with organic and inorganic pollutants: State-of-the-art review. Journal of Environmental Chemical Engineering. 2021;**9**:105095

[81] Trindade ACC, De Andrade Silva F, Kriven WM. Mechanical behavior of K-geopolymers reinforced with silane-coated basalt fibers. Journal of the American Ceramic Society. 2021;**104**:437-447

[82] Korniejenko K, Figiela B, Šimonová H, Kucharczyková B, Guigou MD, Łach M. The influence of fibre pre-treatment on the mechanical properties of the geopolymer composites. MATEC Web of Conferences. 2020;**322**:01012. DOI: 10.1051/matecconf/202032201012

[83] Furtos G, Molnar L, Silaghi-Dumitrescu L, Pascuta P, Korniejenko K. Mechanical and thermal properties of wood fiber reinforced geopolymer composites. Journal of Natural Fibers. 2022;**19**:6676-6691. DOI: 10.1080/15440478.2021.1929655

[84] Wang S, Xue Q, Ma W, Zhao K, Wu Z. Experimental study on mechanical properties of fiber-reinforced and geopolymer-stabilized clay soil. Construction and Building Materials. 2021;**272**:121914. DOI: 10.1016/j.conbuildmat.2020.121914

[85] Korniejenko K, Frączek E, Pytlak E, Adamski M. Mechanical properties of geopolymer composites reinforced with natural Fibers. Procedia Engineering. 2016;**151**:388-393. DOI: 10.1016/j.proeng.2016.07.395

[86] Viscusi G, Barra G, Verdolotti L, Galzerano B, Viscardi M, Gorrasi G. Natural fiber reinforced inorganic foam composites from short hemp bast fibers obtained by mechanical decortation of unretted stems from the wastes of hemp cultivations. Materials Today Proceedings. 2021;**34**:176-179. DOI: 10.1016/j.matpr.2020.02.672

[87] Alomayri T, Shaikh FUA, Low IM. Synthesis and mechanical properties of cotton fabric reinforced geopolymer composites. Composites. Part B, Engineering. 2014;**60**:36-42. DOI: 10.1016/j.compositesb.2013.12.036

[88] Benyahia A, Ghrici M, Choucha S, Omran A. Characterization of fiber reinforced self-consolidating mortars for use in patching damaged concrete. Latin American Journal of Solids and Structures. 2017;**14**:1124-1142

[89] Rashad AM. Insulating and fire-resistant behaviour of metakaolin and fly ash geopolymer mortars. Proceedings of the Institution of Civil Engineers - Construction Materials. 2019;**172**:37-44. DOI: 10.1680/jcoma.17.00016

[90] Kamseu E, Ceron B, Tobias H, Leonelli E, Bignozzi MC, Muscio A, et al. Insulating behavior of metakaolin-based geopolymer materials assess with heat flux meter and laser flash techniques. Journal of Thermal Analysis and Calorimetry. 2012;**108**:1189-1199. DOI: 10.1007/s10973-011-1798-9

[91] Duxson P, Provis JL, Lukey GC, Mallicoat SW, Kriven WM, van Deventer JSJ. Understanding the

relationship between geopolymer composition, microstructure and mechanical properties. Colloids and Surfaces A: Physicochemical and Engineering Aspects. 2005;**269**:47-58. DOI: 10.1016/j.colsurfa.2005.06.060

[92] Duxson P, Lukey GC, van Deventer JSJ. Thermal conductivity of Metakaolin Geopolymers used as a first approximation for determining gel interconnectivity. Industrial and Engineering Chemistry Research. 2006;**45**:7781-7788. DOI: 10.1021/ie060187o

[93] Ye H, Zhang Y, Yu Z, Mu J. Effects of cellulose, hemicellulose, and lignin on the morphology and mechanical properties of metakaolin-based geopolymer. Construction and Building Materials. 2018;**173**:10-16. DOI: 10.1016/j.conbuildmat.2018.04.028

[94] Nazari A, Rohani AF. Alkali-activated geopolymer produced by seeded fly ash and rice husk bark ash. Advances in Cement Research. 2012;**24**:301-309. DOI: 10.1680/adcr.11.00038

[95] Zhang SP, Zong L. Evaluation of relationship between water absorption and durability of concrete materials. Advances in Materials Science and Engineering. 2014;**2014**:e650373

[96] Bonet-Martínez E, García-Cobo P, Pérez-Villarejo L, Castro E, Eliche-Quesada D. Effect of olive-pine bottom ash on properties of geopolymers based on metakaolin. Materials. 2020;**13**:901. DOI: 10.3390/ma13040901

[97] Rehman SKU, Imtiaz L, Aslam F, Khan MK, Haseeb M, Javed MF, et al. Experimental investigation of NaOH and KOH mixture in SCBA-based geopolymer cement composite. Materials. 2020;**13**:3437. DOI: 10.3390/ma13153437

[98] Degefu DM, Liao Z, Berardi U, Labbé G. The dependence of thermophysical and hygroscopic properties of macro-porous geopolymers on Si/Al. Journal of Non-Crystalline Solids. 2022;**582**:121432. DOI: 10.1016/j.jnoncrysol.2022.121432

[99] Su Z, Guo L, Zhang Z, Duan P. Influence of different fibers on properties of thermal insulation composites based on geopolymer blended with glazed hollow bead. Construction and Building Materials. 2019;**203**:525-540. DOI: 10.1016/j. conbuildmat.2019.01.121

[100] Baloyi RB, Ncube S, Moyo M, Nkiwane L, Dzingai P. Analysis of the properties of a glass/sisal/polyester composite. Scientific Reports. 2021;**11**:1-10

[101] Fei W, Yutong Y, Shiping Y. Thermal and moisture performance parameters of high toughness engineered cementitious composite(ECC) with PVA fibers. Journal of Building Engineering. 2021;**43**:102905

[102] Shen H, Li Y, Yao W, Yang S, Yang L, Pan F, et al. Solvent-free cellulose nanocrystal fluids for simultaneous enhancement of mechanical properties, thermal conductivity, moisture permeability and antibacterial properties of polylactic acid fibrous membrane. Composites. Part B, Engineering. 2021;**222**:109042

[103] De Azevedo AGS, Strecker K, Lombardi CT. Produção de geopolímeros à base de metacaulim e cerâmica vermelha. Cerâmica. 2018;**64**:388-396

[104] Haruna S, Mohammed BS, Wahab MMA, Liew MS. Effect of paste aggregate ratio and curing methods on the performance of one-part alkali-activated concrete. Construction and Building Materials. 2020;**261**:120024

[105] Nedeljković M, Visser J, Šavija B, Valcke S, Schlangen E. Use of fine recycled concrete aggregates in concrete: A critical review. Journal of Building Engineering. 2021;**38**:102196

[106] Neuba LM, Pereira Junio RF, Ribeiro MP, Souza AT, Lima ES, Filho FCG, et al. Promising mechanical, thermal, and ballistic properties of novel epoxy composites reinforced with *Cyperus malaccensis* sedge fiber. Polymers. 2020;**12**:1776

[107] Farhan KZ, Johari MAM, Demirboğa R. Impact of fiber reinforcements on properties of geopolymer composites: A review. Journal of Building Engineering. 2021;**44**:102628

[108] Ahmad J, Ali S, Yu T, Sheikh MN, Hadi MNS. Analytical investigation on the load-moment interaction behavior of the FRP reinforced geopolymer concrete filled FRP tube circular columns. Journal of Building Engineering. 2021;**42**:102818

[109] Ahmad J, Yu T, Hadi MNS. Behavior of GFRP bar reinforced geopolymer concrete filled GFRP tube columns under different loading conditions. Structure. 2021;**33**:1633-1644

[110] Sabu A, Karthi L. Experimental study on hybrid fibre reinforced geopolymer concrete. In: Proceedings of the National Conference on Structural Engineering and Construction Management, Angamaly, India. London, England: Springer International Publishing; 14-15 May 2020;**46**:213-220

[111] Maras MM, Kose MM. Mechanical and microstructural properties of polypropylene fiber-reinforced geopolymer composites. Journal of Fiber Science and Technology. 2019;**75**:35-46

[112] Ilamvazhuthi SS, Gopalakrishna GVT. Performance of geopolymer concrete with polypropylene fibres. International Journal of Innovations in Engineering and Technology. 2013;**3**:148-156

[113] Amran M, Fediuk R, Vatin N, Lee YH, Murali G, Ozbakkaloglu T, et al. Fibre-reinforced foamed concretes: A review. Materials. 2020;**13**:4323

[114] Fediuk R. High-strength fibrous concrete of Russian Far East natural materials. In: Proceedings of the IOP Conference Series: Materials Science and Engineering, Tomsk, Russia. Bristol, England: IOP Publishing; Nov 2015;**116**:9-11

[115] Martins ROG, Alvarenga RDCSS, Pedroti LG, De Oliveira AF, Mendes BC, De Azevedo ARG. Assessment of the durability of grout submitted to accelerated carbonation test. Construction and Building Materials. 2018;**159**:261-268

[116] Bindiganavile V, Goncalves JRA, Boluk Y. Crack growth resistance in fibre reinforced geopolymer concrete exposed to sustained extreme temperatures. In: Key Engineering Materials; Trans Tech Publications Ltd.: Stafa-Zurich, Switzerland. Bäch, Switzerland: Scientific. Net; 2016;**711**:511-518. ISBN 9783035710441

[117] Areias IOR, Vieira CMF, Colorado HA, Delaqua GCG, Monteiro SN, Azevedo ARG. Could city sewage sludge be directly used into clay bricks for building construction? A comprehensive case study from Brazil. Journal of Building Engineering. 2020;**31**:101374

[118] Jianqiang Wei CM. Degradation mechanisms of natural fiber in the matrix of cement composites. Cement and Concrete Research. 2015;**73**:1-16

[119] Haddaji Y, Majdoubi H, Mansouri S, Tamraoui Y, El Bouchti M, Manoun B, et al. Effect of synthetic fibers on the properties of geopolymers based on non-heat treated phosphate mine tailing. Materials Chemistry and Physics. 2021;**260**:124147

[120] Marvila MT, Azevedo ARG, Barroso LS, Barbosa MZ, De Brito J. Gypsum plaster using rock waste: A proposal to repair the renderings of historical buildings in Brazil. Construction and Building Materials. 2020;**250**:118786

[121] Azwa ZN, Yousif BF, Manalo AC, Karunasena W. A review on the degradability of polymeric composites based on natural fibres. Materials and Design. 2013;**47**:424-442. DOI: 10.1016/j.matdes.2012.11.025

Chapter 4

Foaming Processes and Properties of Geopolymer Foam Concrete

Dong Guo, Yun-Lin Liu and Lan-Ping Qian

Abstract

Geopolymer foam concrete (GFC) is created by incorporating pores into geopolymer slurry or mortar. The functional and mechanical properties, such as thermal insulation, compressive strength, and density, are influenced by the characteristics of the pores and the properties of their walls. In chemically foamed GFC, foaming agents mixed with the geopolymer slurry release gases through decomposition or reaction. The interaction between the geopolymer and the generated gas during foaming determines the pore characteristics. Gas pressure within the pores is influenced by the gas generation rate of the foaming agent, while confinement pressure is affected by surface tension, viscous resistance, and the weight of the slurry, with viscous resistance being the most significant factor. The foaming process concludes when equilibrium is achieved between gas and confinement pressures or when the slurry solidifies. Various alkali solutions with Na_2O contents ranging from 4–7% and modulus ratios from 1.1 to 1.5 were utilized in the mixture design, and the resulting properties were compared to assess their effects. The GFC's density, compressive strength, and thermal conductivity ranged from 280.8 to 865.8 kg/m^3, 1.10 to 8.13 MPa, and 0.088 to 0.20 W/(m × K), respectively. Test results indicated that a higher gas generation rate and lower slurry viscosity—achieved with increased Na_2O content and decreased modulus ratio—led to greater porosity, reduced compressive strength, and lower thermal conductivity.

Keywords: geopolymer foam concrete (GFC), foaming process, pore characteristics, compressive strength, thermal conductivity

1. Introduction

Foam concrete is extensively utilized in construction due to its advantages over conventional concrete, such as reduced density, improved thermal insulation, and enhanced explosion resistance [1–5]. These benefits arise from the presence of closed pores or voids within the concrete, which are introduced through either chemical or physical methods. Traditionally, ordinary Portland cement (OPC) has been used as the precursor for foam concrete. However, OPC production is associated with approximately 8% of global anthropogenic CO_2 emissions and other environmental pollutants [3, 6]. In response, geopolymer materials have emerged as a more sustainable alternative. Geopolymers offer reduced CO_2 emissions and allow the incorporation of various industrial by-products, including fly ash (FA) and ground granulated blast furnace slag (GGBS), as raw materials [7–9].

Geopolymers are generally made from aluminosilicate precursors, such as metakaolin (MK) or fly ash (FA) [10–19], or from calcium aluminosilicate precursors, like ground granulated blast furnace slag (GGBS) [20–23], or a combination of these. These precursors contain high levels of SiO_2, Al_2O_3, and CaO, and can be activated using alkaline solutions, usually consisting of MOH and $M_2O \cdot nSiO_2$ (where M is Na or K). Upon geopolymerization, these materials form a disordered alkali aluminosilicate gel that encapsulates unreacted precursors and inert minerals [1]. Geopolymer materials with well-designed mixture proportions can exhibit mechanical properties that are comparable to or exceed those of OPC materials [7, 24, 25]. Additionally, geopolymers can be tailored for various strength requirements depending on their mix proportions [7, 26].

Geopolymer foam concrete (GFC) is created by incorporating voids into geopolymer, and is also named as foam/foamed, cellular, or porous geopolymer [7, 27]. Compared to ordinary Portland foam concrete, geopolymers offer higher strength, which contributes to lower density and improved thermal insulation at equivalent compressive strengths [28, 29]. Additionally, the reduced chemically bound water content and more discontinuous gel structures in geopolymer materials further enhance the thermal insulation properties of the geopolymer matrix [23]. Moreover, the inherent advantages of geopolymer, such as corrosion resistance and strong durability [30], are preserved in geopolymer foam concrete [5, 31–33].

Both physical (i.e., mechanical) [16, 17, 22, 23] and chemical methods [10–15, 18–21] are used to introduce pores into geopolymer foam concrete (GFC). Physical methods involve mixing the geopolymer slurry with foam, which is generated by machinery, such as the NCT foam generator. In contrast, the chemical foaming method mixes the slurry with foaming agents, which release gases through decomposition or reactions with the liquid matrix. Hydrogen peroxide (H_2O_2) [10–12, 18, 21] and aluminum (Al) powder [13–15, 19, 20] are commonly used foaming agents. Specifically, H_2O_2 produces oxygen via the reaction $2H_2O_2 \rightarrow 2H_2O + O_2\uparrow$ [34], while Al powder generates hydrogen gas through a reaction with hydroxyl ions: $2Al + 2OH^- + 2H_2O \rightarrow 2AlO_2^- + 3H_2\uparrow$ [35]. Al powder is efficient in gas generation, with 1 g (approximately 0.7 mL) of Al powder generating 1.24 L of hydrogen gas. Al is abundant, inexpensive (primarily due to electrical energy costs), and lightweight [36]. It is also relatively stable under ambient conditions due to its dense oxidation layer, making it easier and safer to store and transport [37]. Moreover, the Al hydroxide by-product is environmentally benign [38] and can be incorporated into geopolymerization. Some researchers have questioned the reliability of chemical foaming methods due to their sensitivity to reaction temperature and curing conditions [39, 40]. However, using Al powder as the foaming agent in geopolymer foam concrete mitigates this issue, given the higher hydroxyl concentration in geopolymer slurry compared to OPC slurry. Additionally, the chemical foaming method tends to produce more regular and reproducible pores compared to physical foaming methods [12, 41].

Geopolymer foam concrete (GFC) is a two-phase material consisting of a geopolymer matrix and voids. Consequently, its functional and mechanical properties (e.g., thermal conductivity, sound absorption, compressive strength, flexural strength) are determined by both the characteristics of matrix (e.g., compressive strength and thermal conductivity) and pores (e.g., porosity, pore size distributions, average pore size, and roundness). Hilal et al. [42] observed a nearly linear relationship between the compressive strength of foam concrete and that of the matrix. Additionally, enhanced thermal insulation in GFC can be attained by utilizing a matrix material with lower thermal conductivity [16]. The properties of the geopolymer matrix are affected by factors such as the precursor type, the admixture of the alkaline solution, and curing temperature [12]. Important characteristics of the alkaline solution include the Na_2O content and the silica modulus. Studies have shown that the compressive strength of geopolymers increases with Na_2O content up to a certain threshold [43, 44]. Furthermore, strength enhancements at higher modulus ratios have been reported [45]. This improvement is attributed to reduced porosity and a more compact microstructure, resulting from a more condensed and polymerized C-S-H gel structure. However, strength enhancement may plateau or decrease when the modulus ratio exceeds a critical value [43, 45–48], which is influenced by the mineralogical composition of the precursors [47] and the Na_2O content [48].

The functional and mechanical properties of GFC are also affected by pore characteristics, such as porosity, average pore size, and shape factor (roundness). Specifically, increased porosity and larger pore sizes typically result in lower density and reduced compressive strength [16, 19, 42, 49]. Moreover, higher porosity is typically associated with lower thermal conductivity [16], though the relationship between pore size and thermal conductivity is less clear. Shao et al. [16] found that smaller pore sizes contribute to lower thermal conductivity, but excessively reducing the pore size can impair thermal insulation performance. Additionally, irregular pore shapes, which may result from squeezing of the geopolymer slurry [49] or the presence of through-holes, negatively affect both the strength and temperature insulation of GFC [16]. The degree of irregularity can be quantified by roundness, which is calculated as the ratio of the perimeter squared to the sectional area multiplied by 4π (i.e., roundness = $\text{perimeter}^2/(4\pi \times \text{Area})$).

Pore characteristics are shaped by the interaction between pores and the geopolymer slurry during the casting/foaming process. The internal forces acting on physically foamed concrete have been analyzed in previous studies [40, 49, 50] to prevent the collapse of GFC. Physically foamed bubbles are subjected to buoyancy, viscous resistance, and gravity (i.e., drainage force [50]). The imbalance between gravity and buoyancy causes bubbles to move vertically, resulting in a nonuniform distribution of bubble size along the height. In extreme cases, bubble volume and buoyancy may increase significantly due to the rupture of slurry walls and coalescence of smaller bubbles, until the bubbles rise to the surface and burst. Excessive gas escape from GFC can cause structural collapse [50, 51], making pore coalescence undesirable in physically foamed GFC. Hao et al. [40] proposed requirements for slurry properties to mitigate this issue.

In addition to discussing the preparation techniques and macroscopic properties of GFC, this chapter also explores the relationship between pore characteristics and macroscopic properties. It further investigates the interaction between binder and chemically produced gas during the foaming process, which at last determines the final pore structures and macroscopic properties of the GFC.

2. Foaming process

Figure 1 schematically demonstrates the forces applied on a single pore in chemically foamed GFC. Unlike physically foamed concrete, the pores in

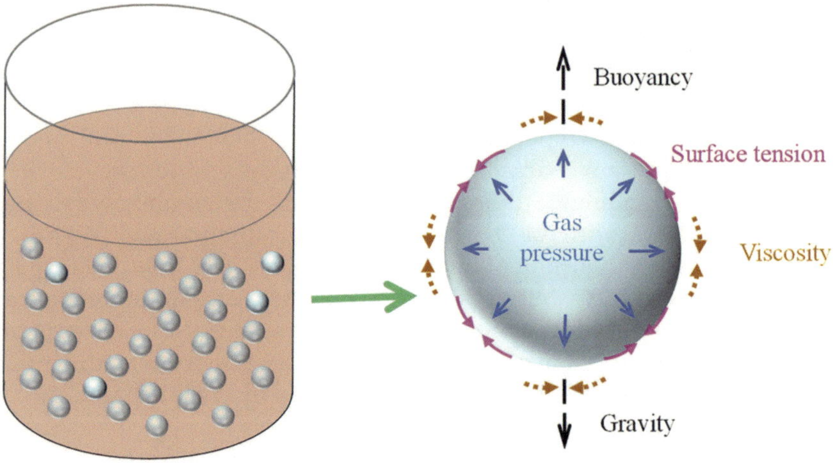

Figure 1.
Force analysis of a single pore during the foaming process.

chemically foamed GFC are generated by gases released from the foaming agent upon mixing with the slurry. The pores within GFC are subject to both internal gas pressure, determined by the extent of the Al powder reaction, and confinement force, which is influenced by the slurry's rheology [40, 51, 52]. Initially, the gas pressure exceeds the confinement force, causing the pore volume to expand. As this occurs, the slurry walls surrounding the pores thin, promoting the coalescence of pores, as depicted in **Figure 2**.

The expansion and coalescence of pores, as illustrated in **Figure 2**, are driven by the internal gas pressure within the pores and the pressure difference between adjacent pores. These processes are restrained by the confinement forces, including surface tension, viscosity resistance, and gravity of the slurry [40, 51, 52]. Among these, viscosity resistance, resulting from the movement of slurry during pore expansion and coalescence, is the dominant factor. The rate of gas formation reduces as the increases of Al powder reaction degree. The rheology of the slurry is influenced by the mixture proportions and the geopolymerization level. The foaming process concludes after achieving the equilibrium between the gas pressure and the surrounding confinement forces, or when the slurry hardens [50].

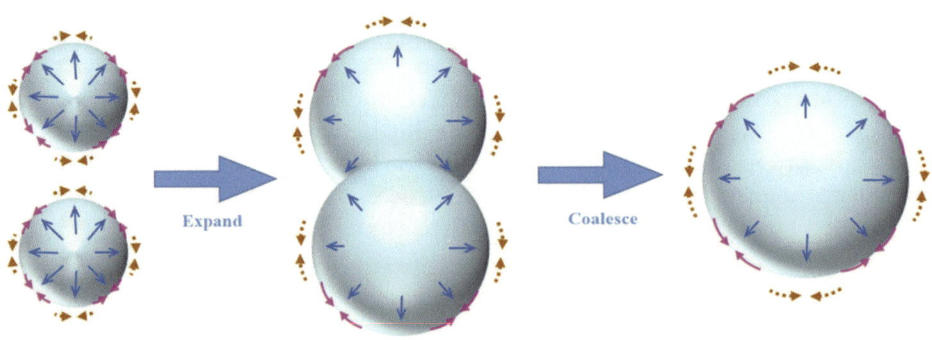

Figure 2.
Expand and coalescence of the pores.

In the former case, the geopolymerization degree of the geopolymer does not change significantly before reaching the aforementioned equilibrium. Then, the foaming process ends when the gas generation rate of the foaming agents reduces. While the gas generation rate can be significantly affected by the concentration of OH- in the geopolymer. Additionally, the rheological properties of geopolymer, including plastic viscosity and yield stress, are interrelated [53] and arise from the flocculation of particles, which involves Walls, electrostatic forces, colloidal, and contact interactions [54, 55]. As reported in previous studies, the rheological properties of geopolymer slurries can be modified by using different alkaline solutions (e.g., Na_2O content [53] and modulus ratio [56]). Additionally, the surface tension of geopolymer can be significantly influenced by the presence of surfactant [50].

In the latter case, the foaming process concludes when the geopolymer matrix reaches sufficient strength, making the efficiency of the foaming process dependent on the balance between the gas formation rate and the binder's reaction rate [57]. If the slurry sets too quickly, the subsequent expansion of pores due to the gas generation is restricted, resulting in lower porosity in the GFC. Conversely, a lower setting rate can lead to ultra-high porosity and potential structural instability [58, 59]. In this case, the variation of rheological properties of the geopolymer should be considered to optimize pore characteristics. In the initial stage, the reaction mainly includes dissolution, gelation, and polymerization. Their reaction rates can be modified by using alkaline activators with different Na_2O contents and modulus ratios [60], as well as different precursor types [7].

Based on the above, the gas generation rates, rheological properties of the geopolymer, and reaction rate determine the foaming kinetics of the GFC and are influenced by the admixture design of the alkaline solution (i.e., Na_2O content and modulus ratio). Therefore, the effects of alkaline solutions on the foaming mechanism, pore characteristics, and macroscopic properties of geopolymer foam concrete will be discussed in the subsequent sections.

3. Experimental program

3.1 Materials

Class F fly ash (FA) from Henan Yulian Energy Co. Ltd. and ground granulated blast furnace slag (GGBS) from Tangshan Iron and Steel Co. Ltd. were used as precursors for the geopolymer matrix. The D50 (median particle diameters) of FA, GGBS, and Al powder were measured as 14.2 μm, 11.1 μm, and 2.68 μm. Additionally, the specific surface areas of FA and GGBS were 2.38 m^2/g and 1.34 m^2/g, respectively. The particle size distributions of FA, GGBS, and Al powder were measured using BET method and are shown in **Figure 3**. The chemical compositions of FA and GGBS were analyzed using X-ray fluorescence (XRF) and are presented in **Table 1**.

The water glass used as the alkaline activator had a modulus ratio of 2.25 and a concentration of 50°Bé, containing 29.9% SiO_2 and 13.75% Na_2O. The modulus of the alkaline activator varies within the range of 1.1 to 1.5. The modulus ratio was adjusted by adding sodium hydroxide (NaOH) flakes with a purity of 98%. Calcium stearate, an anionic surfactant known for its foam-stabilizing properties due to its two hydrophobic non-polar long chains [61, 62], was added. The calcium stearate used had a purity greater than 99.7%.

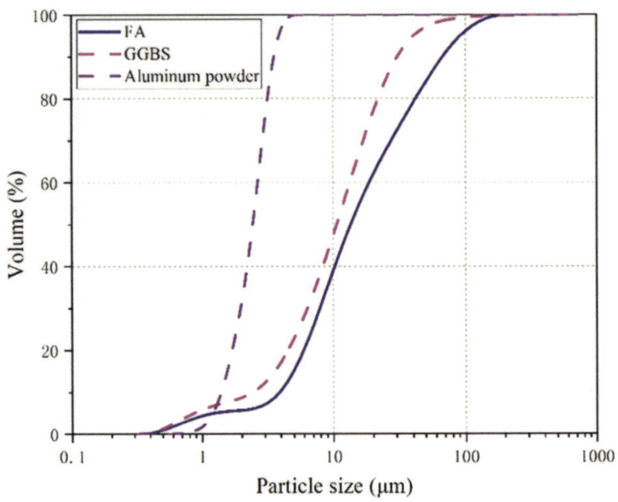

Figure 3.
The cumulative curves of FA, GGBS, and Al powder.

Table 2 presents the mixture compositions of the specimens. In the first specimen series (numbered 1 to 4), the modulus ratio was maintained at 1.5, while the Na_2O contents varies from 4–7%. In the second specimen series (numbered 5 to 9), the Na_2O content was held constant at 5%, while the modulus ratios varied from 1.1 to 1.5.

	SiO_2	Al_2O_3	CaO	Fe_2O_3	K_2O	MgO	TiO_2	SO_3	Na_2O
FA	54.97	33.9	3.38	2.05	1.46	1.11	1.02	0.891	0.556
GGBS	31.24	16.92	39.41	—	0.354	7.97	1.19	1.45	—

Table 1.
Chemical compositions of FA and GGBS by XRF (%).

No.	FA (g)	GGBS (g)	Water (g)	NaOH (g)	Water glass (g)	Al (g)	Calcium stearate (g)	Na_2O content (%)	Modulus ratio
1	30	70	33.30	1.80	19.36	0.30	0.10	4	1.5
2	30	70	31.50	2.25	24.20	0.30	0.10	5	1.5
3	30	70	30.00	2.70	29.04	0.30	0.10	6	1.5
4	30	70	28.30	3.15	33.88	0.30	0.10	7	1.5
5	30	70	34.80	3.39	17.90	0.30	0.10	5	1.1
6	30	70	34.00	3.10	19.50	0.30	0.10	5	1.2
7	30	70	33.20	2.79	21.20	0.30	0.10	5	1.3
8	30	70	32.20	2.54	22.80	0.30	0.10	5	1.4
9(2)	30	70	31.50	2.25	24.20	0.30	0.10	5	1.5

Note: Mixing No. 9(2) means the No. 9 mixture design is identical to the No. 2 mixture design.

Table 2.
Mixture compositions.

3.2 Preparation of specimens

Figure 4 demonstrates the casting procedure of specimens. Initially, NaOH flakes are dissolved in water to create NaOH solution, which is then cooled in an alkali-proof container to mitigate accelerated reactions caused by the exothermic nature of NaOH [63]. To maintain its effectiveness, the NaOH solution should be used within 24 hours. In the next step, the NaOH solution was mixed with a water glass to prepare the alkaline activator. Concurrently, the Al powder suspension is created by mixing Al powder with water.

During the mixing process, the dry ingredients (e.g., FA, GGBS, and calcium stearate) are first blended for 2 minutes. Afterward, the alkaline activator is added and mixed for about 90 seconds. Next, the aluminum powder suspension is poured into the mixer. Finally, the mixture is mixed at a rate of 1200 r/min for 30 seconds and poured into a mold to a depth of 1/2. The outer surface is covered with cling film to prevent cracking caused by moisture evaporation, and the mixture is cured at room temperature for 24 hours. After demoulding, the specimens are transferred to a curing chamber maintained at 20°C and 95% humidity.

3.3 Test methods

The duration of the foaming process was determined based on the volume variation of GFC, which was measured using a graduated cylinder shown in **Figure 5.** After the stirring in mixer, the slurry was immediately poured into the cylinder. The initial volume of GFC was set as 20 mL. Then, the variation of volume was recorded at 10-second intervals.

Cubic GFC specimens (40 mm × 40 mm × 40 mm) were prepared for testing the compressive strength in accordance with ASTM C1329 [64]. Then, the dry density of GFC was calculated by the weight of fully dried specimens dividing their dimensions, as specified in ASTM C1329 [64]. Additionally, cubic specimens with dimensions of 100 mm × 100 mm × 100 mm were prepared for each mixture composition to obtain cross-sectional images and analyze pore characteristics.

Thermal conductivity was tested using 300 mm × 300 mm × 30 mm specimens, following GB/T 10294–2008 [65], with a heat flow meter apparatus (DD300F-D15

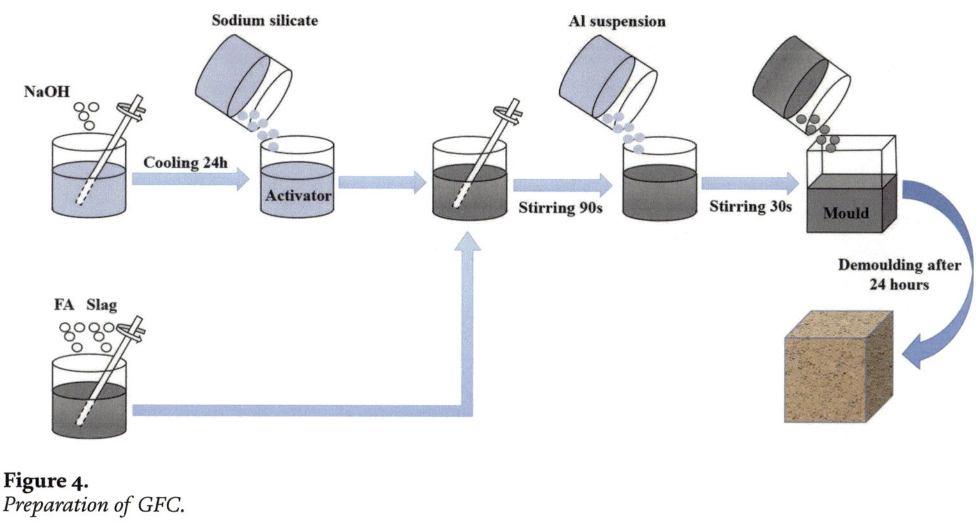

Figure 4.
Preparation of GFC.

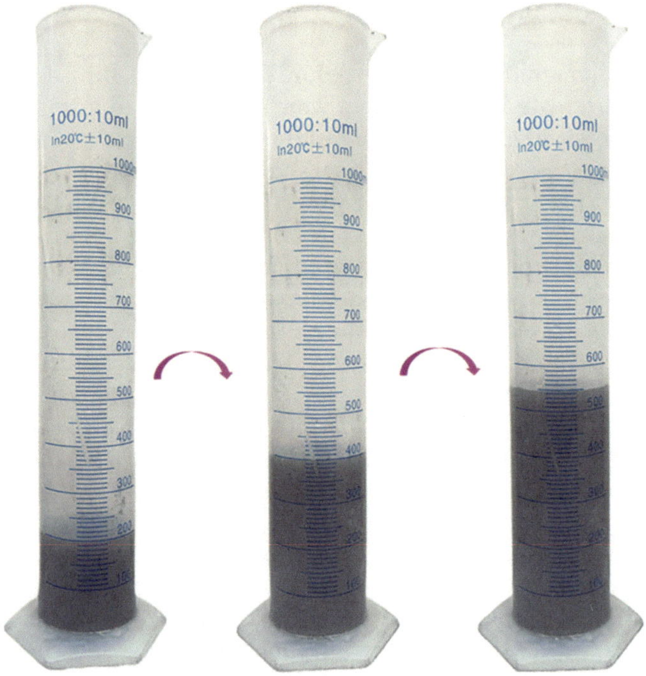

Figure 5.
Measurement of GFC volume.

from Foreda, China). During the test, the hot plate (35°C) and cold plate (15°C) were placed on opposite surfaces of the specimen. The thermal conductivity was measured in a steady state after approximately 2 hours of testing.

Pore structures of the GFC specimens were tested on fully cured specimens after drying at 60°C for one day. The samples were vertically cut from the pouring surface to the bottom using a saw. The cross-section images were then captured using a 3R-MSUSB401 (Anyty, Japan) optical microscope, as shown in **Figure 6a**. The roundness and size

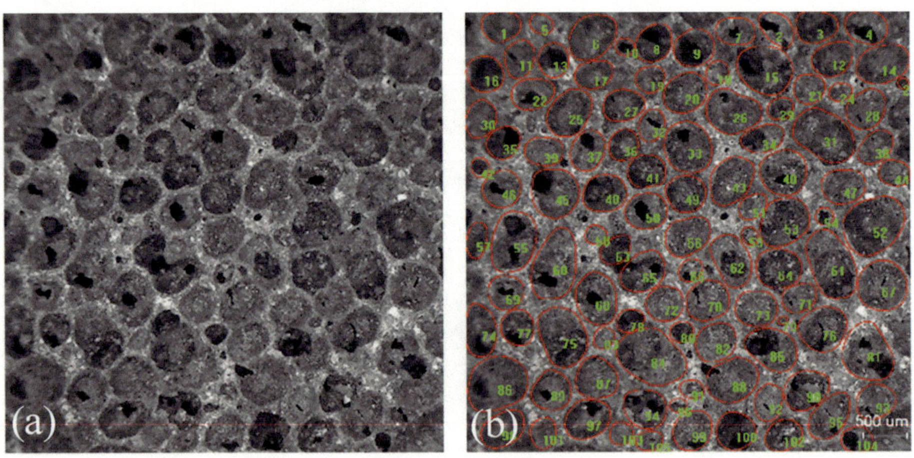

Figure 6.
Microscopic images of GFC: (a) before processing; and (b) after processing.

distribution of the pores were then determined using image analysis software (Image Pro Plus 6.0) [21, 22]. **Figure 6b** shows the processed cross-section image. In this study, the smallest discernible pore size was 6.8 μm, which is comparable to the objective macroscopic scale pores (approximately 5.6 μm) in other studies on GFC [57].

Figure 7 shows the NDJ-5S rotational viscometer (Roweeltec, Shanghai, China), which was adopted for measuring the viscosity of the slurry. The tests were conducted with a 3# stirrer (covering a maximum measurement range from 0 to 20,000 MPa·s) and at a speed of 6 rpm. Each test required three minutes, and viscosity variations during the foaming process were recorded at five-minute intervals.

During the geopolymerization reaction, hydrogen was generated from the reaction between aluminum powder and the alkaline activator. The volume of hydrogen

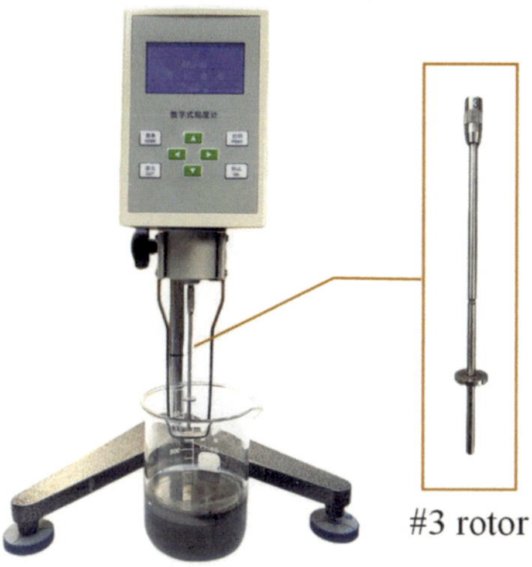

Figure 7.
Rotational viscometer.

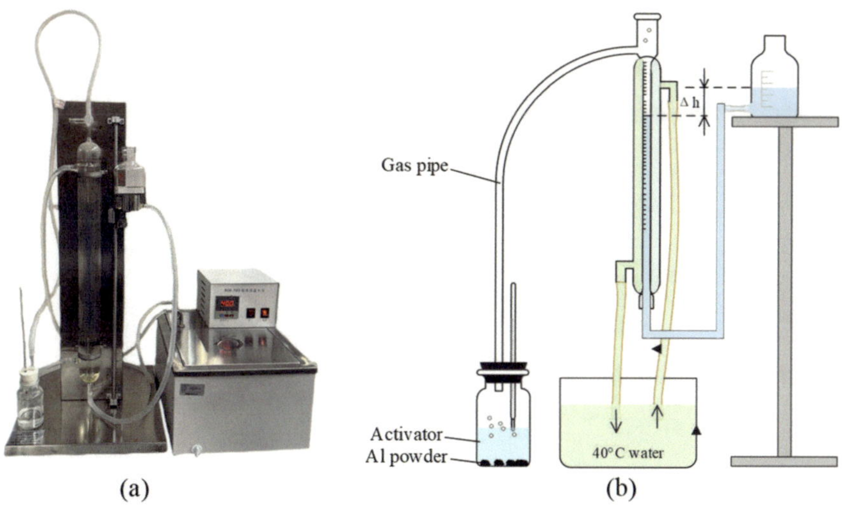

Figure 8.
Gas generation meter (a) photo; and (b) schematic.

was measured using a gas generation meter (Onway, Shanghai, China), as shown in **Figure 8**, in accordance with JC/T 407–2008 [66]. The gas was produced in a sealed beaker and transferred to the water level meter (a slim measuring cylinder) through gas pipes. To prevent the high temperature of the hydrogen gas from affecting the measuring water, a constant-temperature water tank was used.

Before testing, the measuring water was added to the water level meter, and the measuring cylinder was placed at the same level. The hydrogen gas was generated once the Al powder was added to the sealed beaker. The gas in the pipe then pushes the meter water (shown as blue liquid in the figure) down in the slim measuring cylinder and up in the stout measuring cylinder. Ultimately, the volume of hydrogen was determined by the change in water levels in the slim measuring cylinder. During this test, the temperatures of the hydrogen and meter water were maintained at 40°C using temperature-controlled water (green liquid in the figure).

4. Results and discussion

4.1 Effect of the Na_2O content

Figure 9 shows cross-sectional photos of the GFC with Na_2O contents of 4% and 7%, respectively. The dimensions of the section are 100 mm × 100 mm. The images reveal that the pores are spherical and uniformly distributed throughout the height of the specimen, indicating excellent foam quality. Additionally, the pore sizes are noticeably larger at higher Na_2O content compared to lower Na_2O content.

4.1.1 Effect of Na_2O content during the foaming process

Figure 10 illustrates the change in GFC volume during the foaming process, with volumes normalized to the initial 20 mL. The volume of GFC increased continuously throughout the foaming process and gradually reached a maximum value. It is evident

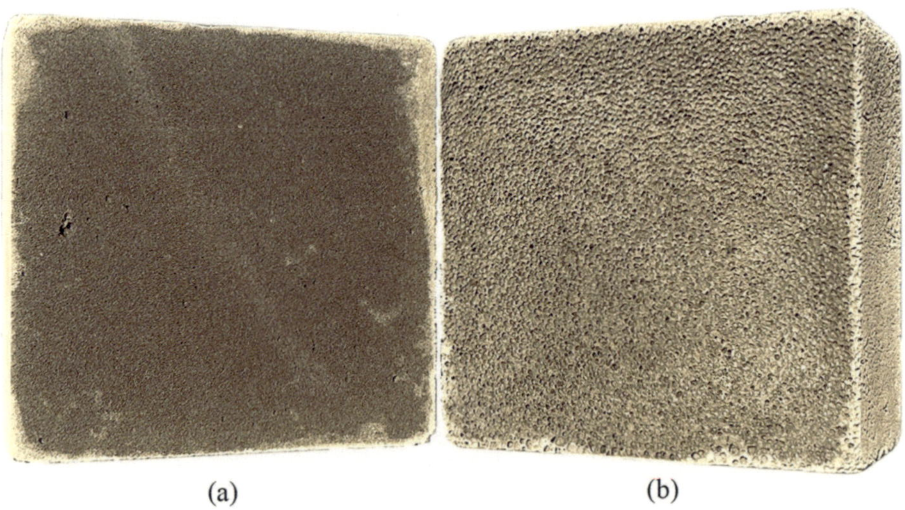

Figure 9.
Cross-section images of the GFC with Na_2O contents of: (a) 4%; and (b) 7%.

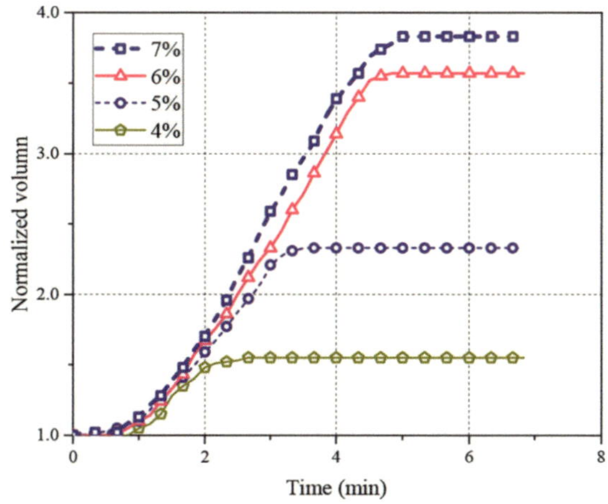

Figure 10.
The volume increase history of GFC at different Na_2O contents.

that at higher Na_2O contents, the foaming durations were generally longer, resulting in higher final volumes. This indicates that gas generation efficiency improved with higher Na_2O content. Specifically, the final volume of GFC with 7% Na_2O content (766 mL) was approximately 2.5 times greater than that with 4% Na_2O content (310 mL).

Figure 11 shows the variations in gas volume measured by the gas meter (**Figure 8**) at different Na_2O contents. The gas volume increase rate at higher Na_2O content (7%) was slightly higher than at lower Na_2O content (4%), likely because Al particles had a greater chance of reacting with free OH^-. Additionally, the difference in the final gas volume at varying Na_2O contents was much smaller than the difference observed in GFC volumes. According to Bernoulli's principle, the air pressure inside the pores at higher Na_2O content should be higher, which can be attributed to the differing rheological properties of the slurry.

Figure 12 shows the viscosity variations of the geopolymer slurry over 20 minutes after mixing, at different Na_2O contents. The viscosities of most slurries remained

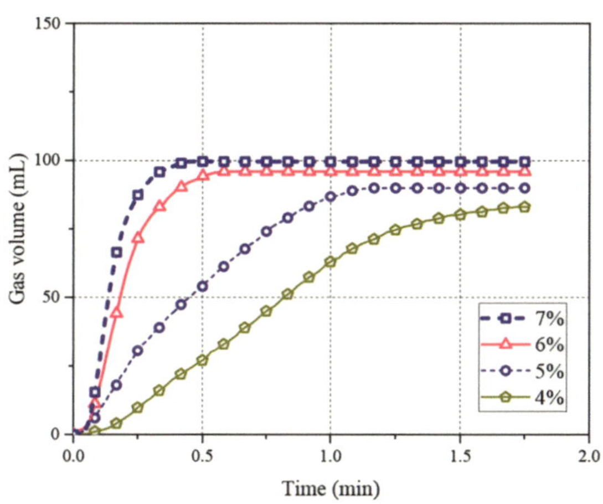

Figure 11.
The gas volume increase history at different Na_2O contents.

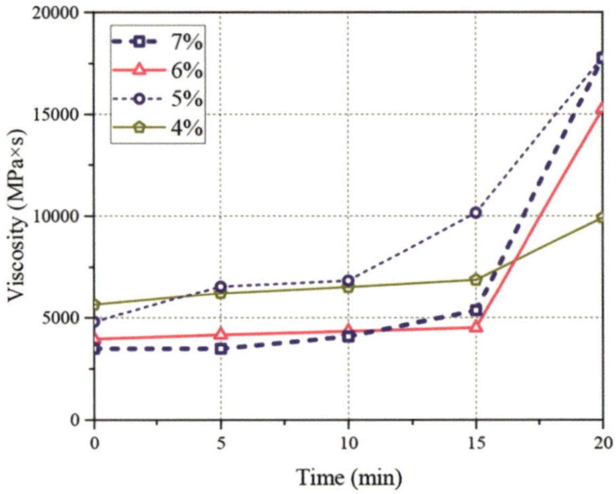

Figure 12.
Variation of viscosity of geopolymer slurry at different Na_2O contents.

stable and low during the first 15 minutes, facilitating pipe transportation [58]. However, viscosities increased significantly from 15 to 20 minutes. When viscosity exceeded 10,000 MPa·s, the slurry's fluidity was greatly reduced, making further foaming impossible. According to **Figures 10** and **12**, foaming processes were generally completed within 5 minutes of mixing, with slurries maintaining relatively invariant viscosities. Therefore, the foaming process was ended after the balance between gas pressure and confinement pressure was achieved (as discussed in the introduction), rather than the setting of the slurry.

During the first 15 minutes, the slurry with higher Na_2O content exhibited lower viscosity. This can be attributed to the higher dispersion of geopolymer grains, such as amorphous silica and alumina [43, 53]. The increased Na_2O content created a higher pH environment, which enhanced the dissolution of FA and GGBS grains [67]. The greater dissolution of aluminate and silicate altered particle geometry and Zeta potential, reducing particle flocculation and improving the rheological properties of the geopolymer slurry [53].

Additionally, the viscosity of geopolymer slurry decreased from 5799 MPa·s to 3949 MPa·s as Na_2O content increased from 5–6%. However, the decrease rate was lower (11.8%, from 3949 MPa·s to 3482 MPa·s) as Na_2O content rose from 6–7%. This lower decrease rate can be attributed to the limited dissolution of silica and alumina from FA due to the precipitation of geopolymer gels on FA particle surfaces [43].

Moreover, the dissolution of silica and alumina from the raw materials is more extensive at higher Na_2O content, which accelerated geopolymer synthesis and increased the amount of synthesis products [43, 68] and led to a significant viscosity increase after the initial 15 minutes. While the specimens at lower Na_2O content, i.e., 3%, were also prepared but exhibited significantly lower early-stage (1 day) strengths due to incomplete geopolymerization, resulting in demoulding failures.

4.1.2 Effect of Na_2O content on the properties of GFC

Figure 13 shows the pore characteristics of GFC at different Na_2O contents. At 4% Na_2O content, most pores (75%) were in the range of 200 to 400 μm, with an average pore size of approximately 328 μm. In contrast, at 7% Na_2O content, 69% of the pores

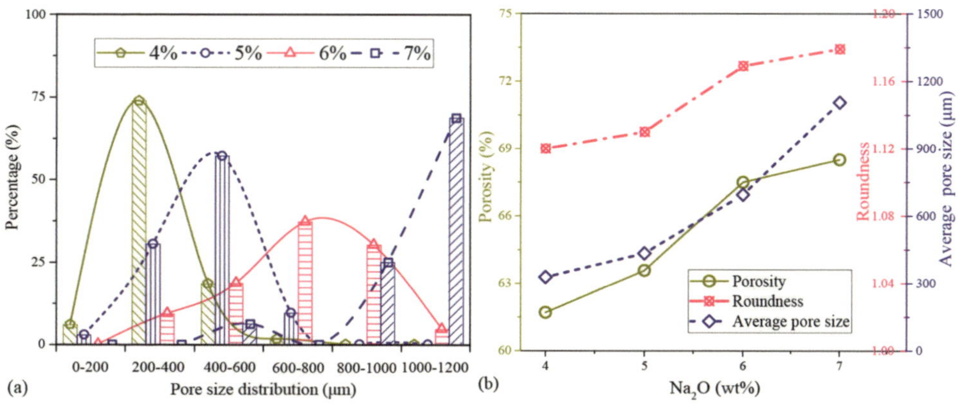

Figure 13.
Pore characteristics: (a) pore size distribution; and (b) porosity, average pore size, and roundness of GFC at different Na_2O contents.

were larger than 1000 μm, with an average pore size of 1105 μm. As shown in **Figure 13b**, both porosity and average pore size increased with Na_2O content. This enhancement in pore expansion and coalescence during the foaming process is attributed to the lower slurry viscosity (**Figure 12**) and higher gas generation rate (**Figure 11**) [51, 52]. Additionally, higher Na_2O content resulted in increased pore roundness, indicating more regular (spherical) pore shapes. Previous studies have shown that pore roundness is related to porosity, with higher porosity leading to greater pore merging and irregularity [49]. Compared to physically foamed concrete using a similar approach, chemically foamed concrete exhibited smaller roundness magnitudes even with larger average pore sizes [49], demonstrating superior pore quality.

Figure 14 illustrates the variation in the properties of hardened GFC at different Na_2O contents. It can be seen that GFC with higher porosity and larger average pore sizes exhibited lower dry density and thermal conductivity, which can be traced back to the increased porosity and average pore size at higher Na_2O content. In addition, as Na_2O content increased from 4–7%, the dry density and thermal conductivity decreased by 61.7% and 56%, respectively, as shown in **Figure 14a**.

Previous studies have shown that compressive strength in geopolymer paste tends to either increase monotonically or first increase and then decrease with rising Na_2O content within a specific range [43, 69], primarily due to higher reaction rates of FA and enhanced formation of N-A-S-H gels [70]. However, in the case of GFC specimens, the compressive strength decreased consistently as Na_2O content increased from 4–7%, dropping from 8.13 MPa to 1.10 MPa at 28 days (**Figure 14b**). These results suggest that pore characteristics significantly influence the compressive strength of GFC. The higher compressive strength of GFC can be achieved at thicker geopolymer paste walls between pores, which was achieved at lower Na_2O content, higher slurry viscosity (**Figure 12**), and lower gas generation rate (**Figure 11**) [16].

4.2 Effect of modulus ratio

Figure 15 compares the cross-sectional images of GFC specimens with different modulus ratios but identical Na_2O contents. It is evident that the pore dimensions and porosity are significantly higher at a lower modulus ratio (1.1, shown in **Figure 15a**) compared to a higher modulus ratio (1.5, shown in **Figure 15b**).

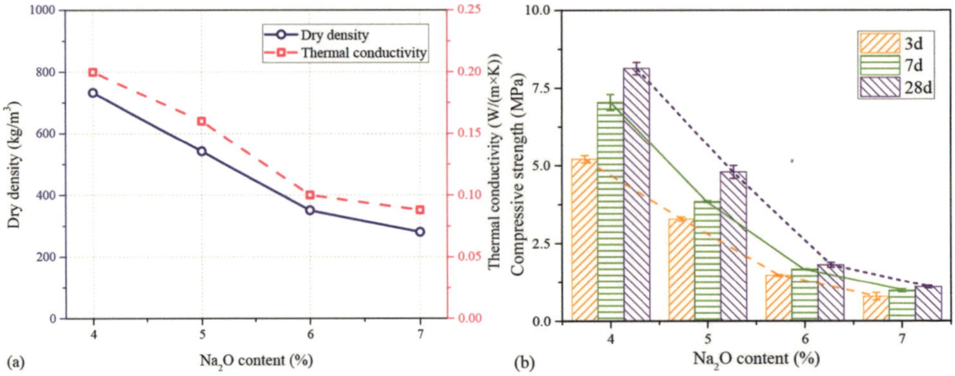

Figure 14.
Effect of Na_2O content on the (a) density and thermal conductivity; and (b) compressive strength of GFC.

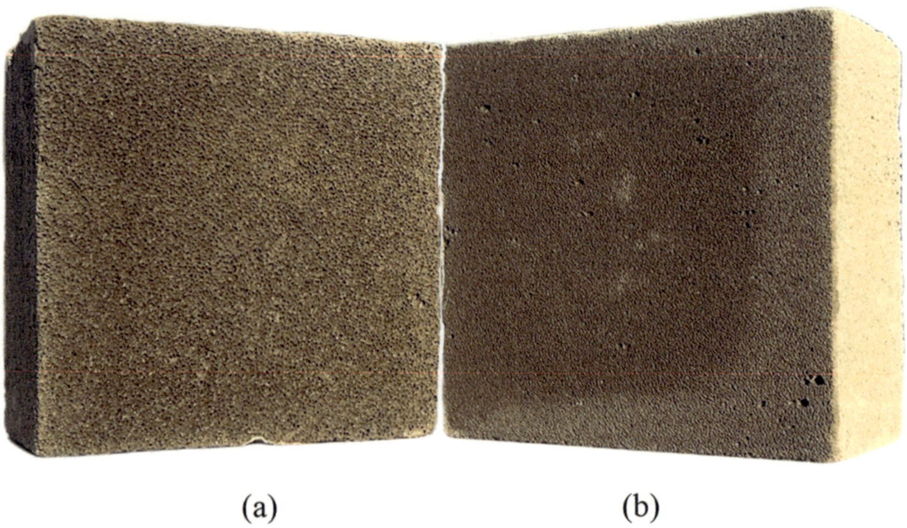

Figure 15.
Cross-section images of the GFC specimens with modulus ratios of (a) 1.1; and (b) 1.5.

4.2.1 Effect of modulus ratio during the foaming process

Figure 16 illustrates the variations in GFC volume during the foaming process. It can be generally seen that the quicker volume increase rate and higher volume of GFC were achieved at a lower modulus ratio. For instance, the GFC with a modulus ratio of 1.1 achieved its ultimate volume in 2.0 minutes, while the GFC with a modulus ratio of 1.5 took about 3.3 minutes to reach its ultimate volume. Additionally, the final volume of GFC with a modulus ratio of 1.1 was about 1.33 times greater than that with a modulus ratio of 1.5. These differences in increase rates and final volumes can be attributed to variations in gas generation rates and the rheological properties of the geopolymer slurry.

Figure 17 shows the measured variations in gas volume using the gas meter (**Figure 8**) at different modulus ratios. It is clear that the alkaline activator with a modulus ratio of 1.1 facilitated a significantly higher gas generation rate than that

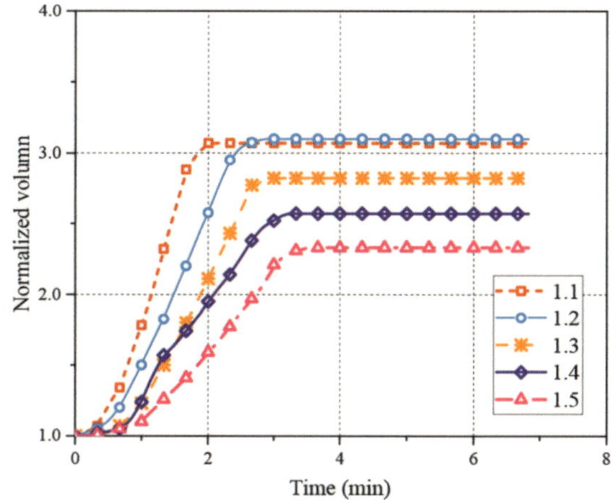

Figure 16.
Variation of GFC volume at different modulus ratios.

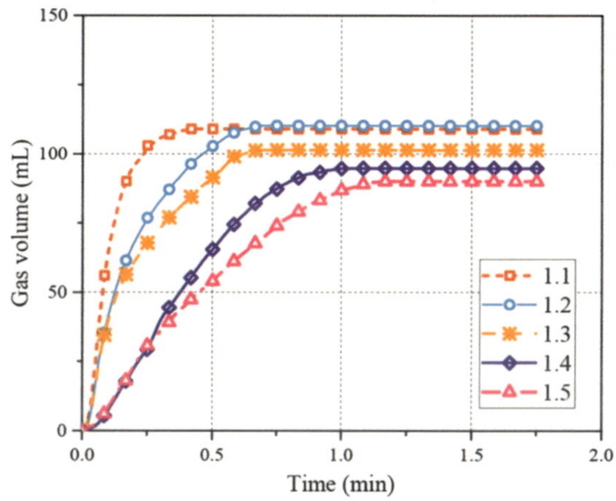

Figure 17.
Variations of gas volumes at different modulus ratios.

at a modulus ratio of 1.5, along with a final gas volume that was 1.21 times greater (109 mL compared to 90 mL).

Figure 18 shows the variations in slurry viscosities at different modulus ratios. Similar to the results observed at different Na_2O contents (**Figure 12**), the slurry viscosities increased at a slower rate within the first 15 minutes after stirring. Additionally, **Figure 18** indicates that the viscosity of the slurry was higher at higher modulus ratios. This phenomenon can be attributed to the increased content of silicate ions in the slurry at higher modulus ratios, which enhanced the reactions with calcium ions and facilitated the formation of primary C-S-H. The C-S-H deposits among the GGBS grains increased interactions between the grains, resulting in higher viscosity [71].

4.2.2 Effect of modulus ratio on properties of GFC

Figure 19 shows the resultant pore characteristics at different modulus ratios. The GFC with modulus ratios ranging from 1.2 to 1.4 exhibited dimensions of pores

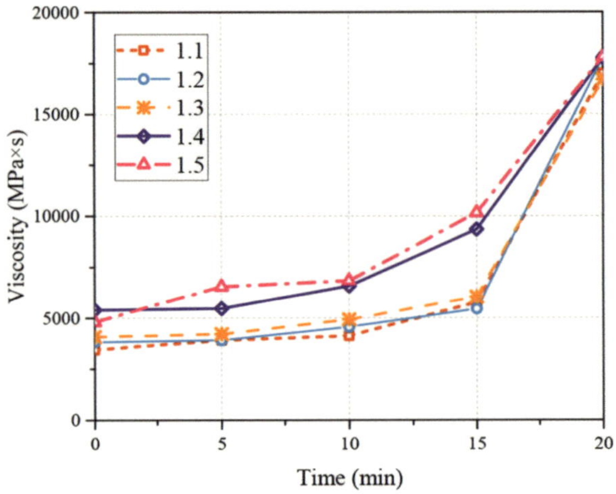

Figure 18.
Variations of GFC viscosities at different modulus ratios.

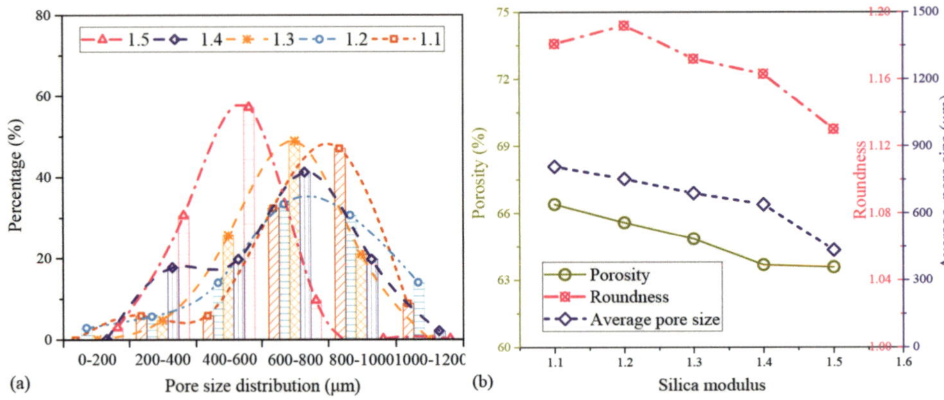

Figure 19.
Pore characteristics of GFC: (a) pore size distribution; and (b) porosity, average pore size, and roundness at different modulus ratios.

ranging from 600 to 800 μm. In contrast, the GFC with a modulus ratio of 1.5 exhibited up to 57% smaller pores (400 to 600 μm). Additionally, as the modulus ratio increased from 1.1 to 1.5, the average pore size and porosity of GFC decreased by 4.1% and 21.0%, respectively (**Figure 19b**). The reduced pore size and pore roundness at higher modulus ratios can be attributed to the enhanced slurry viscosity (**Figure 18**) and reduced gas generation rate of aluminum powder (**Figure 17**).

Figure 20 illustrates the effect of modulus ratio on the properties of hardened GFC. The density of GFC increased from 358.8 kg/m^3 at a modulus ratio of 1.1 to 542.1 kg/m^3 at a modulus ratio of 1.5. Similarly, the thermal conductivity rose from 0.115 W/(m·K) to 0.16 W/(m·K) over the same modulus ratios. In addition, the compressive strength of GFC increased from 2.43 MPa to 4.79 MPa, as the modulus ratio increased from 1.1 to 1.5 (**Figure 20b**). Previous studies have reported increased compressive strength in geopolymer pastes with higher modulus ratios due to increased soluble Si ion concentrations facilitating greater silicate

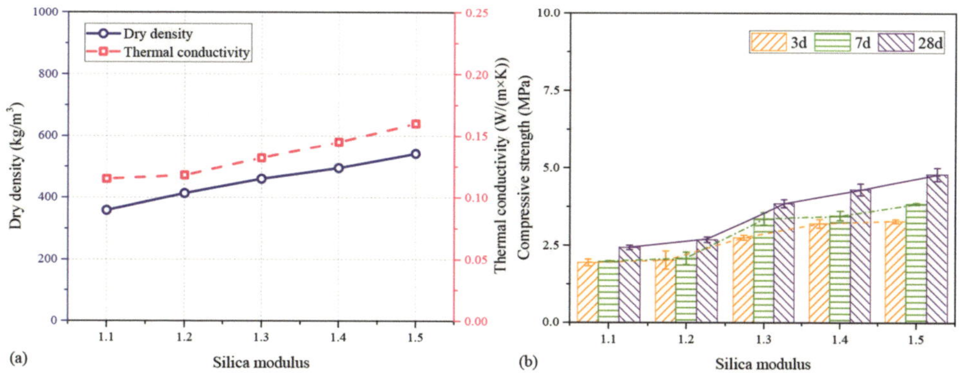

Figure 20.
The variation of properties of GFC: (a). thermal conductivity and dry density; and (b). compressive strength at various silica modulus.

polymerization [45]. However, the increased strength should be primarily attributed to the thicker barriers between pores inside GFC, which is due to the lower viscosity of geopolymer.

5. Conclusions

This paper presents the preparation and characterization of geopolymer foam concrete (GFC). It examines the variation of rheological properties and the gas generation rate during the foaming process at different Na_2O contents and modulus ratios. Subsequently, the resultant pore characteristics and macroscopic properties were analyzed. Based on the experimental findings, the following conclusions can be drawn:

1. Geopolymer foam concretes with Na_2O contents ranging from 4–7% and modulus ratios between 1.1 and 1.5 were successfully prepared, demonstrating dry densities from 280.8 kg/m^3 to 865.8 kg/m^3 and compressive strengths between 1.10 MPa and 8.13 MPa.

2. The variation in macroscopic properties of GFC, including dry density, thermal conductivity, and compressive strength, can be traced back to the variation in the pore characteristics.

3. Pore characteristics are influenced by the gas generation rate and the rheological properties of the geopolymer slurry during the foaming stage. GFC with higher porosity and larger average pore sizes can be produced at lower geopolymer viscosity and higher gas generation rates.

4. Higher Na_2O content and lower modulus ratios result in lower thermal conductivity and compressive strength of GFC, attributed to increased gas formation rates and lower slurry viscosity.

5. The effect of the activator on the compressive strength of GFC is primarily due to changes in pore characteristics rather than the strength of the geopolymer matrix.

Acknowledgements

The authors would like to acknowledge the financial support received from National Natural Science Foundation of China (52408323; 52178278); the Department of Science and Technology of Guangdong Province, China (2023A1515110842); the University Synergy Innovation Program of Anhui Province (GXXT-2023-061); the China Postdoctoral Science Foundation (2024M750152); the open Foundation of the Key lab of Prefabricated Building Research Institute of Anhui Province (AHZPY2023KF02).

Conflict of interest

The authors declare no conflict of interest.

Author details

Dong Guo[1]*, Yun-Lin Liu[2,3] and Lan-Ping Qian[4]

1 School of Civil Engineering and Transportation, Guangzhou University, Guangzhou, People's Republic of China

2 Prefabricated Building Research Institute of Anhui Province, Anhui Jianzhu University, Hefei, People's Republic of China

3 School of Civil Engineering, Anhui Jianzhu University, Hefei, People's Republic of China

4 Faculty of Architecture, Civil and Transportation Engineering, Beijing University of Technology, Beijing, People's Republic of China

*Address all correspondence to: dong.gzhu@gzhu.edu.cn

References

[1] Ramamurthy K, Nambiar EK, Indu S. A classification of studies on properties of foam concrete. Cement and Concrete Composites. 2009;**31**(6):388-396

[2] Terzi A, Pezo L, Miti V, Radojevi Z. Artificial fly ash based aggregates properties influence on lightweight concrete performances. Ceramics International. 2015;**41**(2):2714-2726

[3] Sang G, Zhu Y, Yang G, Zhang H. Preparation and characterization of high porosity cement-based foam material. Construction and Building Materials. 2015;**91**:133-137

[4] Ding Z, Dong B, Xing F, Han N, Li Z. Cementing mechanism of potassium phosphate based magnesium phosphate cement. Ceramics International. 2012;**38**(8):6281-6288

[5] Bildirici ME. Cement production, environmental pollution, and economic growth: Evidence from China and USA. Clean Technologies and Environmental Policy. 2019;**21**(4):783-793

[6] Provis JL. Geopolymers and other alkali activated materials: Why, how, and what? Materials and Structures. 2014;**47**(1-2):11-25

[7] Shang J, Dai JG, Zhao TJ, Guo SY, Zhang P, Mu B. Alternation of traditional cement mortars using fly ash-based geopolymer mortars modified by slag. Journal of Cleaner Production. 2018;**203**(PT.1-1216):746-756

[8] Qian L-P, Xu L-Y, Alrefaei Y, Wang T, Ishida T, Dai J-G. Artificial alkali-activated aggregates developed from wastes and by-products: A state-of-the-art review. Resources, Conservation and Recycling. 2022;**177**:105971

[9] Ahmad MR, Qian L-P, Fang Y, Wang A, Dai J-G. A multiscale study on gel composition of hybrid alkali-activated materials partially utilizing air pollution control residue as an activator. Cement and Concrete Composites. 2023;**136**:104856

[10] Abdollahnejad Z, Pacheco-Torgal F, Félix T, Tahri W, Aguiar JB. Mix design, properties and cost analysis of fly ash-based geopolymer foam. Construction and Building Materials. 2015;**80**:18-30

[11] Palmero P, Formia A, Antonaci P, Brini S, Tulliani J-M. Geopolymer technology for application-oriented dense and lightened materials. Elaboration and characterization. Ceramics International. 2015;**41**(10):12967-12979

[12] Ducman V, Korat L. Characterization of geopolymer fly-ash based foams obtained with the addition of Al powder or H2O2 as foaming agents. Materials Characterization. 2016;**113**:207-213

[13] Hlaváček P, Šmilauer V, Škvára F, Kopecký L, Šulc R. Inorganic foams made from alkali-activated fly ash: Mechanical, chemical and physical properties. Journal of the European Ceramic Society. 2015;**35**(2):703-709

[14] Arellano Aguilar R, Burciaga Díaz O, Escalante García JI. Lightweight concretes of activated metakaolin-fly ash binders, with blast furnace slag aggregates. Construction and Building Materials. 2010;**24**(7):1166-1175

[15] Novais RM, Ascensão G, Ferreira N, Seabra MP, Labrincha JA. Influence of water and aluminium powder content on the properties of waste-containing

geopolymer foams. Ceramics International. 2018;**44**(6):6242-6249

[16] Shao N, Zhang Y, Liu Z, Wang D, Zhang Z. Fabrication of hollow microspheres filled fly ash based foam geopolymers with ultra-low thermal conductivity and relative high strength. Construction and Building Materials. 2018;**185**:567-573

[17] Abdullah MMAB, Hussin K, Bnhussain M, Ismail KN, Yahya Z, Razak RA. Fly ash-based geopolymer lightweight concrete using foaming agent. International Journal of Molecular Sciences. 2012;**13**(6):7186-7198

[18] Shi J, Liu B, Liu Y, Wang E, He Z, Xu H, et al. Preparation and characterization of lightweight aggregate foamed geopolymer concretes aerated using hydrogen peroxide. Construction and Building Materials. 2020;**256**:119442

[19] Dembovska L, Bajare D, Ducman V, Korat L, Bumanis G. The use of different by-products in the production of lightweight alkali activated building materials. Construction and Building Materials. 2017;**135**:315-322

[20] Esmaily H, Nuranian H. Non-autoclaved high strength cellular concrete from alkali activated slag. Construction and Building Materials. 2012;**26**(1):200-206

[21] Xu F, Gu G, Zhang W, Wang H, Huang X, Zhu J. Pore structure analysis and properties evaluations of fly ash-based geopolymer foams by chemical foaming method. Ceramics International. 2018;**44**(16):19989-19997

[22] Wang L, Tan X. Preparation and properties of alkali activated foam cement reinforced with polypropylene fibers, journal of Wuhan University of Technology-mater. Science Education. 2011;**26**(5):960-964

[23] Zhang Z, Provis JL, Reid A, Wang H. Mechanical, thermal insulation, thermal resistance and acoustic absorption properties of geopolymer foam concrete. Cement and Concrete Composites. 2015;**62**:97-105

[24] Ding Y, Dai JG, Shi CJ. Mechanical properties of alkali-activated concrete: A state-of-the-art review. Construction and Building Materials. 2016;**127**(nov.30):68-79

[25] Yang YL, Jia JZ, Zhang MY, Dai GJ. Effects of nano-TiO_2 on strength, shrinkage and microstructure of alkali activated slag pastes. Cement and Concrete Composites. 2015;**57**:1-7

[26] Lao J-C, Huang B-T, Fang Y, Xu L-Y, Dai J-G, Shah SP. Strain-hardening alkali-activated fly ash/slag composites with ultra-high compressive strength and ultra-high tensile ductility. Cement and Concrete Research. 2023;**165**:107075

[27] Ding Y, Dai JG, Shi CJ. Fracture properties of alkali-activated slag and ordinary Portland cement concrete and mortar. Construction and Building Materials. 2018;**165**:310-320

[28] Pasupathy K, Ramakrishnan S, Sanjayan J. Formulating eco-friendly geopolymer foam concrete by alkali-activation of ground brick waste. Journal of Cleaner Production. 2021;**325**:129180

[29] Fang Y, Ahmad MR, Lao J-C, Qian L-P, Dai J-G. Development of artificial geopolymer aggregates with thermal energy storage capacity. Cement and Concrete Composites. 2023;**135**:104834

[30] Zheng H, Dai JG, Poon CS, Li W. Influence of calcium ion

in concrete pore solution on the passivation of galvanized steel bars. Cement and Concrete Research. 2018;**108**:46-58

[31] Masi G, Rickard W, Vickers L, Bignozzi MC, Riessen AV. A comparison between different foaming methods for the synthesis of light weight geopolymers. Ceramics International. 2014;**40**(9):13891-13902

[32] Xiong Y, Zhu Y, Chen C, Zhang Y. Effect of nano-alumina modified foaming agents on properties of foamed concrete. Construction and Building Materials. 2020;**267**:121045

[33] Wang X, Huang J, Dai S, Ma B, Jiang Q. Investigation of silica fume as foam cell stabilizer for foamed concrete. Construction and Building Materials. 2020;**237**:117514

[34] Huang Z, Zhang T, Wen Z. Proportioning and characterization of Portland cement-based ultra-lightweight foam concretes. Construction and Building Materials. 2015;**79**:390-396

[35] Hilal AA, Thom NH, Dawson AR. On entrained pore size distribution of foamed concrete. Construction and Building Materials. 2015;**75**:227-233

[36] Czech E, Troczynski T. Hydrogen generation through massive corrosion of deformed aluminum in water. International Journal of Hydrogen Energy. 2010;**35**(3):1029-1037

[37] Padró CEG, Lau F. Advances in Hydrogen Energy. Berlin: Springer; 2000

[38] Rosenband V, Gany A. Application of activated aluminum powder for generation of hydrogen from water. International Journal of Hydrogen Energy. 2010;**35**(20):10898-10904

[39] Liu Y, Chen Y, Deng J, Wang J. N-doped aluminum-graphite (Al-gr-N) composite for enhancing in-situ production and activation of hydrogen peroxide to treat landfill leachate. Applied Catalysis B: Environmental. 2021;**297**:120407

[40] Hao Y, Yang G, Liang K. Development of fly ash and slag based high-strength alkali-activated foam concrete. Cement and Concrete Composites. 2022;**128**:104447

[41] Zhang Z, Provis JL, Reid A, Wang H. Geopolymer foam concrete: An emerging material for sustainable construction. Construction and Building Materials. 2014;**56**:113-127

[42] Hilal AA, Thom NH, Dawson AR. On void structure and strength of foamed concrete made without/with additives. Construction and Building Materials. 2015;**85**:157-164

[43] Cho Y-K, Yoo S-W, Jung S-H, Lee K-M, Kwon S-J. Effect of Na_2O content, SiO_2/Na_2O molar ratio, and curing conditions on the compressive strength of FA-based geopolymer. Construction and Building Materials. 2017;**145**:253-260

[44] Qian L-P, Wang Y-S, Alrefaei Y, Dai J-G. Experimental study on full-volume fly ash geopolymer mortars: Sintered fly ash versus sand as fine aggregates. Journal of Cleaner Production. 2020;**263**:121445

[45] Cihangir F, Ercikdi B, Kesimal A, Ocak S, Akyol Y. Effect of sodium-silicate activated slag at different silicate modulus on the strength and microstructural properties of full and coarse sulphidic tailings paste backfill. Construction and Building Materials. 2018;**185**:555-566

[46] Criado M, Fernández-Jiménez A, De La Torre A, Aranda M, Palomo A. An XRD study of the effect of the SiO_2/Na_2O ratio on the alkali activation of fly ash. Cement and Concrete Research. 2007;**37**(5):671-679

[47] Firdous R, Stephan D. Effect of silica modulus on the geopolymerization activity of natural pozzolans. Construction and Building Materials. 2019;**219**:31-43

[48] Sivasakthi M, Jeyalakshmi R. Effect of change in the silica modulus of sodium silicate solution on the microstructure of fly ash geopolymers. Journal of Building Engineering. 2021;**44**:102939

[49] Gu G, Xu F, Ruan S, Huang X, Zhu J, Peng C. Influence of precast foam on the pore structure and properties of fly ash-based geopolymer foams. Construction and Building Materials. 2020;**256**:119410

[50] Jones MR, Ozlutas K, Zheng L. Stability and instability of foamed concrete. Magazine of Concrete Research. 2016;**68**(11):542-549

[51] Medri V, Papa E, Dedecek J, Jirglova H, Benito P, Vaccari A, et al. Effect of metallic Si addition on polymerization degree of in situ foamed alkali-aluminosilicates. Ceramics International. 2013;**39**(7):7657-7668

[52] Roviello G, Menna C, Tarallo O, Ricciotti L, Messina F, Ferone C, et al. Lightweight geopolymer-based hybrid materials. Composites Part B: Engineering. 2017;**128**:225-237

[53] Zhang D-W, Wang D-M, Liu Z, Xie F-Z. Rheology, agglomerate structure, and particle shape of fresh geopolymer pastes with different NaOH activators content. Construction and Building Materials. 2018;**187**:674-680

[54] Roussel N, Lemaître A, Flatt RJ, Coussot P. Steady state flow of cement suspensions: A micromechanical state of the art. Cement and Concrete Research. 2010;**40**(1):77-84

[55] Yim HJ, Kim JH, Shah SP. Cement particle flocculation and breakage monitoring under Couette flow. Cement and Concrete Research. 2013;**53**:36-43

[56] Rodrigue Kaze C, Adesina A, Alomayri T, Assaedi H, Kamseu E, Chinje Melo U, et al. Characterization, reactivity and rheological behaviour of metakaolin and meta-halloysite based geopolymer binders. Cleaner Materials. 2021;**2**:100025

[57] Cui Y, Wang D, Zhao J, Li D, Ng S, Rui Y. Effect of calcium stearate based foam stabilizer on pore characteristics and thermal conductivity of geopolymer foam material. Journal of Building Engineering. 2018;**20**:21-29

[58] Dhasindrakrishna K, Pasupathy K, Ramakrishnan S, Sanjayan J. Progress, current thinking and challenges in geopolymer foam concrete technology. Cement and Concrete Composites. 2021;**116**:103886

[59] Medri V, Ruffini A. Alkali-bonded SiC based foams. Journal of the European Ceramic Society. 2012;**32**(9): 1907-1913

[60] Kumarappa DB, Peethamparan S, Ngami M. Autogenous shrinkage of alkali activated slag mortars: Basic mechanisms and mitigation methods. Cement and Concrete Research. 2018;**109**:1-9

[61] Sawicki GC. Impact of surfactant composition and surfactant structure on foam control performance. Colloids and Surfaces A: Physicochemical

and Engineering Aspects. 2005;**263**(1-3):226-232

[62] Wu J, Zhang Z, Yi L, Dongxu. Preparation and characterization of ultra-lightweight foamed geopolymer (UFG) based on fly ash-metakaolin blends. Construction and Building Materials. 2018;**168**:771-779

[63] Görhan G, Kürklü G. The influence of the NaOH solution on the properties of the fly ash-based geopolymer mortar cured at different temperatures. Composites Part B: Engineering. 2014;**58**:371-377

[64] ASTM, C1329/C1329M-16a. Standard Specification for Mortar Cement. West Conshohocken, PA: ASTM International; 2016

[65] GB/T, 10294-2008. Thermal Insulation Determination of Steady-State Thermal Resistance and Related Properties Guarded Hot Plate Apparatus. Beijing: Standards Press of China; 2008

[66] JC/T, 407-2008. Aluminium Pastes for Aerated Concrete. Beijing: Standards Press of China; 2008

[67] Deir E, Gebregziabiher BS, Peethamparan S. Influence of starting material on the early age hydration kinetics, microstructure and composition of binding gel in alkali activated binder systems. Cement and Concrete Composites. 2014;**48**:108-117

[68] Rifaai Y, Yahia A, Mostafa A, Aggoun S, Kadri E-H. Rheology of fly ash-based geopolymer: Effect of NaOH concentration. Construction and Building Materials. 2019;**223**:583-594

[69] Fernández-Jiménez A, Palomo A. Composition and microstructure of alkali activated fly ash binder: Effect of the activator. Cement and Concrete Research. 2005;**35**(10):1984-1992

[70] Zhang J, Shi C, Zhang Z. Effect of Na_2O concentration and water/binder ratio on carbonation of alkali-activated slag/fly ash cements. Construction and Building Materials. 2021;**269**:121258

[71] Qureshi MN, Ghosh S. Effect of silicate content on the properties of alkali-activated blast furnace slag paste. Arabian Journal for Science and Engineering. 2014;**39**(8):5905-5916

Chapter 5

Perspective Chapter: Performance of Geopolymer-Based Coating for Protection of Concrete Surface after Coupling Effect of Wetting-Drying Cycles and Seawater

Pavlo Kryvenko, Igor Rudenko and Oleksandr Konstantynovskyi

Abstract

In marine structures, the risk of degradation of both the concrete structure and the initiation of corrosion of steel reinforcement is precisely caused by the periodicity of wetting in seawater and drying under the influence of atmospheric CO_2. To protect reinforced concrete from the influence of the mentioned ions under coupling exposure classes XS3 and XC4, a coating based on geopolymer of the formula $(0{,}2K_2O + 0{,}8Na_2O)\cdot 4{,}5SiO_2\cdot Al_2O_3\cdot nH_2O$ was designed. The effect of the wastes and industrial by-products as the modifying additives on the acceleration of the synthesis of water-resistant zeolite-like matrices of geopolymer at normal temperatures ($t = 20 \pm 2°C$) was studied. The water resistance coefficient of the geopolymer-based mortars was the highest (> 0.9) for both the slaked lime ($Ca(OH)_2$) and aluminum fluoride (AlF_3) modifications. The effectiveness of geopolymer-based coatings was confirmed by acceptable changes in water absorption (≤ 1 kg/m^2) and adhesion ($\leq$ 23%) after 50 cycles and 200 cycles of wetting in seawater and further drying at $t = 105 \pm 2°C$. Diffusion permeability of Cl^-, SO_4^{2-}-ions into the coatings was to a depth of no more than 1...2 mm, and the coating's thickness $\geq$ 2 mm was determined as appropriate value.

Keywords: aggressive ions, binding, coating, diffusion, permeability, geopolymer, modifying additive, seawater, water resistance, wetting-drying cycles

1. Introduction

To increase the durability of building structures, especially under the effect of aggressive environments, is current global trend in modern materials science [1, 2]. Reinforced concrete structures operating in seawater (berths, piers, coast-protecting structures, dams, etc.) require special attention to their durability [3]. This is caused by the degradation of concrete and the corrosion of steel reinforcement under the action of chlorides (NaCl, $MgCl_2$, $CaCl_2$, and KCl) and sulfates ($MgSO_4$ and $CaSO_4$), the main constituents of seawater [4].

1.1 The effect of seawater

The interaction of SO_4^{2-}-ions with portlandite $Ca(OH)_2$ and calcium hydroaluminates leads to the formation of gypsum $CaSO_4 \cdot 2H_2O$ and secondary ettringite $3CaO \cdot Al_2O_3 \cdot 3CaSO_4 \cdot 32H_2O$ [5]. The increased chloride content in the pore solution causes the formation of phases like calcium oxychlorides $xCa(OH)_2 \cdot yCaCl_2 \cdot zH_2O$ through the reaction with portlandite [6]. The crystallization pressure of these phases in micropores can result in concrete degradation [5, 7].

Moreover, the formation of magnesium hydrosilicate M-S-H is possible as a result of a cation exchange reaction with $MgCl_2$, replacing Ca^{2+}-ion with Mg^{2+}-ion in calcium hydrosilicates [8]. The decalcification of hydrosilicates causes the degradation of cement stone.

Aggressive seawater ions can also initiate corrosion of steel reinforcement [9]. Thus, Cl^--ions cause pitting corrosion of steel rebar [10, 11]. Sulfate ions do not cause immediate depassivation of steel reinforcement but determine the formation of hydrogen sulfide H_2S, which catalyzes the oxidation (carbonation) of hydrate phases [12].

1.2 Effect of exposure zones

The marine structures are subject to different exposure classes [13]:

- XS1 (exposed to airborne salt but not in direct contact with seawater)—atmospheric zone.
- XS2 (permanently submerged)—submerged zone.
- the combination of exposure classes XS3 (periodical effect of seawater or air containing aerosols of seawater) and XC4 (effect of carbon dioxide while periodical contact with water)—splash and tidal zones.

In the *atmospheric zone* of marine structures, corrosion of steel reinforcement caused by airborne chloride is the prevailing degradation mechanism compared to one caused by carbonization [14]. The risk of deterioration of reinforced concrete structures is lowest at exposure class XS1 and can occur due to salt crystallization.

In the *submerged zone*, the permeability of the concrete is determined by diffusion, and is the most important performance than that of the strength [15]. The degradation of concrete under exposure class XS2 is caused by individual and combined effects of chemicals, mainly chlorides and sulfates, leaching of portlandite $Ca(OH)_2$, etc. [4].

Reinforced concrete structures in *splash zone* and *tidal zone* are exposed to the most aggressive effect with combination of exposure classes XC4 and XS3. The transport of Cl^--ions into the concrete occurs under cyclic wetting in seawater, and drying under the influence of atmospheric CO_2 (further, wetting-drying cycles) is higher than the full immersion in submerged zone [16]. An increase in the transport of Cl^--ions into concrete under wetting-drying cycles is due to the following processes:

(1) sorption under wetting; (2) diffusion after evaporation of water under drying; and (3) capillary absorption under subsequent wetting [17].

Wetting-drying cycles also accelerate the transport of $SO4^{2-}$-ions into concrete through capillary adsorption and diffusion [4]. At the same time, sulfate salts are crystallized like needle-like crystals of thenardite Na_2SO_4 and mirabilite $Na_2SO_4 \cdot 10H_2O$ [18]. These phases exert pressure in the micropores, causing concrete degradation [19].

Wetting-drying cycles also increase the destructive effect caused by the carbonation. Accelerated capillary absorption of CO_3^{2-}-ions under wetting-drying cycles and their direct penetration into the concrete during the drying period caused this phenomenon [20].

The combined effect of seawater chlorides and carbonation on reinforced concrete causes the complex of physical and chemical processes [21]. The carbonation can be accelerated in the presence of chlorides [22]. In turn, carbonation promotes the release of Cl^--ions into the pore solution, which have already been bound by gel phases (chemisorption) and layered double hydroxides [23, 24]. This effect causes increased transport of Cl^--ions into the concrete.

There is therefore a need for effective means of protecting marine structures used under wetting-drying cycles from the transport of aggressive ions (Cl^-, SO_4^{2-}, and CO_3^{2-}).

1.3 The protection of marine structures

Nowadays, there is a row of means to limit the transport of aggressive ions from aggressive environment into concrete.

The means for the primary protection of concrete are well known. For example, densification of the concrete structure using supplementary materials like ground granulated blast furnace slag, fly ash, microsilica, etc. [25, 26] or surfactants [27] reduces the permeability. The binding of Cl^- and SO_4^{2}--ions in synthesized *AFm*-phases (i.e., Friedel's salt $3CaO \cdot Al_2O_3 \cdot CaCl_2(SO_4) \cdot 10H_2O$ and Kuzel's salt $3CaO \cdot Al_2O_3 \cdot 0,5CaCl_2 \cdot 0,5SO_4 \cdot 10H_2O$) is another mean to limit their transport [28].

However, the secondary (surface) protection of concrete is a more cost-effective solution for already constructed structures [29]. Depending upon its chemical nature, protective coatings can be organic and inorganic [30]. The main disadvantage of organic coatings applied in a thin layer (100...400 microns) is the decrease in service life under the action of ultraviolet radiation [31].

Portland cement modified with polymer additives (acrylic, epoxy, or polyurethane) is widespread among inorganic coatings [32]. Compared to organic coatings, such coatings have higher performance properties and greater resistance to ultraviolet radiation, but they are very expensive [33].

1.4 Coatings based on geopolymer

Inorganic coatings based on geopolymers (further, geopolymer-based coatings) to protect concrete from the effects of aggressive environments are well known [34, 35]. The increased durability and high protective performances of geopolymer-based coatings are due to the formation of analogs to natural minerals such as zeolites and feldspathoids in the reaction products [36]. To limit the transport of aggressive ions due to the binding by formed zeolite-like matrices allows us to predict the efficiency of geopolymer-based coatings to prevent concrete degradation and to ensure the passive state of the steel reinforcement.

The main limiting factor in the use of geopolymer-based protective coatings is the need to provide elevated temperatures (>40°C) for their hardening to obtain a water-resistant artificial stone with the phases represented by zeolite-like minerals [37]. The synthesis of water-resistant zeolite-like phases (gismondine $CaSi_2Al_2O_8{\cdot}4H_2O$, thompsonite $NaSi_5Al_5O_{20}{\cdot}6H_2O$, and heulandite $Na(K)_6Si_{27}Al_{36}O_{72}{\cdot}24H_2O$) in the reaction products of geopolymer-based coating at normal temperatures (t = 20 ± 2°C) was shown due to optimizing the ratio of oxides in the Na_2O-K_2O-Al_2O_3-SiO_2-H_2O system and application of calcium-containing modifiers (lime, Portland cement, ground granulated blast furnace slag) [38]. The geopolymer-based coating is characterized by high protective performances against sulfate environments [39]. The mitigation of the negative effect of these environments on concrete has been explained by chemisorption and chemical binding of aggressive SO_4^{2-} ions by the zeolite-like phases such as nosean $Na_8(Al_6Si_6O_{24})(SO_4){\cdot}H_2O$, biachellaite $(Na,Ca,K)_8(Si_6Al_6O_{24})(SO_4)_2(OH)_{0.5}{\cdot}H_2O$, and cancrinite $(Na,Ca)_8(Al_6Si_6)O_{24}(CO_3,SO_4)_2{\cdot}2H_2O$, with an increased crystallinity [38].

The above-mentioned water-resistant zeolite-like matrices can also bind Cl^- and CO_3^{2-}-ions. The formation of zeolite-like phases, such as sodalite $Na_8(Al_6Si_6O_{24})Cl_2$, sarcolite $Na_4Ca_{12}Al_8Si_{12}O_{46}(SiO_4,PO_4)(OH,H_2O)_4(CO_3,Cl)$, chabazite $(Al_2CaCl_2O_8Si_2)$, and marinellite $Na_{42}Ca_6Al_{36}Si_{36}O_{144}(SO_4)_8Cl_2{\cdot}6H_2O$, is due to the binding of these aggressive ions [40, 41].

1.5 Choice of modifying additives

A promising area is the search for modifying additives that can contribute to the synthesis of water-resistant zeolite-like matrices of geopolymer-based coatings at normal temperatures as an alternative to Ca-containing additives. These substances can be used either by themselves or in the composition of by-products and production wastes.

The feasibility of incorporation of metals (i.e., Al^{3+}, Fe^{3+}, and Mg^{2+}) into the zeolite framework has been shown [42, 43]. This phenomenon makes it advisable to choose substances containing metals as effective modifiers: a by-product of phosphatic fertilizers production (further, AlF_3 production waste) (content of $AlF_3{\cdot}3.5H_2O$—32.05% by mass, $SiO_2{\cdot}nH_2O$—60.54% by mass) [44]; Fe-containing waste from galvanic wastewater treatment by ferritization method (content of Fe_2O_3— ≥ 30.0% by mass) [45, 46]; zeolitic by-product of the oil industry (content Al_2O_3— ≤ 50.0% by mass) [47]; red mud, by-product of technology to purify bauxite and produce alumina (content Al_2O_3 + Fe_2O_3— ≥ 60.0% by mass) [48].

Salts are another type of substance that can be an effective modifying additive of geopolymer-based coatings. The ability of zeolites to incorporate SO_4^{2-}, Cl^-, CO_3^{2-}, and PO_4^{2-}-ions as a result of exchange reactions explains this thesis [38, 40, 41].

1.6 Estimation of durability

The subject of the discussions is to estimate the durability of geopolymer-based coatings under the action of the mentioned aggressive factors. It is well known that methods of accelerated carbonation are used to determine the atmospheric CO_2 action in the experimentally available period of time [49]. However, there is a difference in the estimation of accelerated and natural carbonation for alkali-activated materials [50]. Furthermore, in the case of accelerated carbonation, it is difficult to simulate drying-wetting cycles that actually affect the structures [51].

The absence of portlandite and the high alkalinity of the solution make the hydration products of alkali-activated materials extremely sensitive to accelerated carbonization [52]. Changes in natrone ($Na_2CO_3{\cdot}10H_2O$)/nahcolite ($NaHCO_3$) equilibria under accelerated carbonization can also lead to an increase in CO_2 permeability due to the lower molecular volume of sodium bicarbonate compared to carbonate [53].

The above results can predict the efficiency of the geopolymer-based coating for improved protection of reinforced concrete against the transport of aggressive ions under wetting-drying cycles. The acceleration of the formation of the water-resistant zeolite-like matrices by the use of modifying additives can result in the limitation of the transport of aggressive ions (Cl^-, SO_4^{2-}, and CO_3^{2-}) in concrete due to their chemisorption and chemical binding. A realistic assessment of the proposed decision requires an objective method to evaluate the aggressive effect of the environment with a combination of exposure classes XS3 and XC4 on reinforcement structure.

Therefore, the aim of the paper is to evaluate the efficiency of concrete protection under wetting-drying cycles by the coating based on geopolymer with the formula $(0.2K_2O + 0.8Na_2O){\cdot}4.5SiO_2{\cdot}Al_2O_3{\cdot}nH_2O$ through changes in performances due to salts, metal oxide and zeolite as modifying additives, which cause the synthesis of water resistance zeolite-like matrices at the normal temperatures.

2. Materials and testing techniques

2.1 Raw materials and mix design

Geopolymer with the following oxide composition $(0.8Na_2O + 0.2K_2O){\cdot}Al_2O_3{\cdot}4.5SiO_2{\cdot}nH_2O$ was used in the design of the coatings.

The major components of the geopolymer were:

- Metakaolin ARGICAL-M 1200S (Imerys Minerals Refractory, France) per CAS 15123–81-6.
- Sodium water glass (TM Klebrig, Ukraine) per CAS 1344-09-8.
- The following components were used to adjust the composition of the geopolymer regarding the content of the main oxides:
- Tripoli powder (PJSC "Kirovograd mechanical plant", Ukraine) per CAS 1317-95-9.
- Solution of potassium hydroxide (KOH) (TM Klebrig, Ukraine) per CAS 1310-58-3.

The composition of the geopolymer was, % by mass: metakaolin – 31.0, sodium water glass – 63.2, tripoli powder – 2.8, potassium hydroxide (calculated on dry matter) – 3.0.

The additives were used as modifiers, affecting the strength and water resistance of the geopolymer at normal temperatures ($t = 20 \pm 2°C$):

- Slaked lime ($Ca(OH)_2$) (TM Klebrig, Ukraine) per CAS 1305-62-0 as an example of *Ca-containing additives.*
- Red mud (Ukraine).
- Aluminum fluoride (AlF_3) per CAS 7784-18-1 as a model of *AlF_3 production waste*.
- Natural zeolite (clinoptilolite) (TM "ZEOLITE-BIO") as a model of *zeolitic by-product*.
- sodium chloride (NaCl) is an example of *sodium salts of strong acids* per CAS 7647-14-5.
- Magnesium oxide (MgO) per CAS 1309-48-4 as an example of *metal oxides*.

A complex additive consisting of sodium phosphate ($Na_3PO_4{\cdot}12H_2O$) per CAS 7601-54-9 (TM Klebrig, Ukraine) and sodium gluconate ($NaC_6H_{11}O_7$) per CAS 527–07-1 (TM Klebrig, Ukraine), was used to provide required consistency and workability retention.

Hydroxyethyl cellulose Tylose 10,000 YP2 (TM Tylose, Germany) per CAS 9004-62-0 was used to add water-retaining capacity to the geopolymer-based coating mix.

Styrene butadiene copolymer powder Axilate PSB 150 (PLC "Synthomer", United Kingdom) was used to increase the bond strength of the coating to concrete substrate.

Polysiloxanes Silsan 1300 (TM Silsan, Germany) was used to increase the water-repellent properties of the coating surface.

Polypropylene fiber (TM MicroArm, Ukraine) was used as a reinforcing agent to increase the crack resistance and physico-mechanical properties of the coatings, length = 2 mm.

Starch ether Opagel FP-9 (TM Opagel, Germany) was used as additive to improve technological properties (consistency, thixotropy, viscosity, etc.).

CEN Standard sand per EN 196–1 was used as aggregate of geopolymer-based mortars for determination of their efficiency.

The fillers used in geopolymer-based coating were:

- fly ash (per the National Standard of Ukraine DSTU B V.2.7–205);
- siliceous sand, fr. 0.16...0.315 mm.

A chemical composition of geopolymer-based coating components is given in **Table 1**. The properties of components in the solid state are given in **Table 2**, and those of the components in the liquid state in **Table 3**.

The mix compositions of geopolymer-based mortar (Mix 1 and Mix 2) for determining the efficiency of modifying additives are given in **Table 4**. Modifying additives were added to Mix 1 in equal quantities—1.7% by mass.

The mix compositions of geopolymer-based coating depending on the type of modifying additive (Mix 3 and Mix 4) are given in **Table 5**. The compositions of geopolymer-based mortars and geopolymer-based coatings used in the study were obtained by improving their efficiency under stationary exposure to sulfate environment (concentration of the sulfate ions 30,000 mg/l) in Ref. [39].

Component	CaO	SiO_2	Al_2O_3	Fe_2O_3	MgO	SO_3	TiO_2	P_2O_5	$K_2O + N_2O$	LOI
Metakaolin	0.27	54.08	43.61	0.77	0.52	—	—	—	0.25	0.50
Tripoli powder	0.83	85.59	6.20	3.15	0.95	0.39	2.03	—	0.67	0.19
Natural zeolite	3.6	73.7	13.1	1.2	0.73	—	0.2	0.11	5.03	2.33
Red mud	10.4	19.58	17.5	37.9	0.20	0.3	6.23	0.19	7.7	—
Fly ash	2.94	52.37	25.25	13.62	2.04	0.41	0.97	—	0.71	1.69

Table 1.
Chemical composition of geopolymer-based coating components (% by mass).

Component	Specific surface area, m^2/kg (by Blaine)	Bulk density, kg/m^3	Specific gravity, g/cm^3	Average particle size, μm	Purity, %
Metakaolin	900	2450	2,4	1.5	93,8
Tripoli powder	800	2300	2,3	3.0	98,0
Slaked lime	400	1230	2,2	5.0	95,0
Natural zeolite	280	1050	2.0	50.0	97.0
Red mud	1500	2650	2.8	1.0	95.0
Fly ash	300	1150	2,1	10.0	97.0

Table 2.
Properties of components in the solid state.

Component	Silicate modulus (M_s)	Density, kg/m^3	Concentration, g/L	Purity, %
Sodium water glass	2.8	1430	574.7	99.0
Solution of potassium hydroxide	—	1430	615.8	95.0

Table 3.
Properties of components in the liquid state.

Component	Mix	
	Mix 1 (without modified additive)	Mix 2
Metakaolin	10.4	9.8
Sodium water glass	21.0	20.0
Tripoli powder	0.9	0.9
Potassium hydroxide KOH (calculated on dry matter)	1.0	0.9
Modifying additive	—	1.7
Siliceous sand per EN 196–1	66.7	66.7

Table 4.
Mix design of geopolymer-based mortars (% by mass).

Seawater of a composition simulating 99.9% of the salts contained in seawater (predominantly, near-neutral salts; % by mass: NaCl – 78.70, $MgCl_2$–9.80, $MgSO_4$–5.76, $CaSO_4$–3.75, KCl – 1.73, $CaCO_3$–0.29) was used [54]. The total concentration of the salts in the seawater was 35 g/l (salinity 35.0‰).

2.2 Testing protocol

The liquid component of geopolymer was prepared by mixing sodium water glass and water solution of KOH. The powder component of geopolymer was prepared by mixing metakaolin, tripoli powder, and modifying additives. The geopolymer was

Component	**Mix**	
	Mix 3	**Mix 4**
Metakaolin	11.91	11.60
Sodium water glass	24.23	23.58
Tripoli powder	1.07	3.65
Potassium hydroxide KOH (calculated on dry matter)	1.14	1.11
Slaked lime	0.77	—
Aluminum fluoride	—	0.77
Sodium phosphate	0.97	0.95
Sodium gluconate	0.42	0.41
Hydroxyethyl cellulose Tylose 10,000 YP2	0.08	0.08
Styrene butadiene copolymer powder Axilate PSB 150	0.51	0.50
Polysiloxanes Silsan 1300	0.71	0.69
Starch ether Opagel FP-9	0.02	0.02
Polypropylene fiber	0.04	0.04
Fly ash	13.82	13.46
Siliceous sand, fr. 0.16...0.315 mm	44.31	43.14

Remarks. The content of tripoli powder in Mix 4 was calculated taking into account the SiO_2 content in AlF_3 production waste [44] (the ratio of tripoli powder to AlF_3 was 2:1).

Table 5.
Mix design of geopolymer-based coatings (% by mass).

prepared by mixing an alkaline solution with a powder component in a Hobart mixer for 3 min at a speed of 140 min^{-1}. The consistency of geopolymer-based mortar mixes and geopolymer-based coating mixes was measured with a Suttard viscometer.

A schematic diagram showing the preparation procedure of geopolymer-based mortar mixes is given in **Figure 1**.

After preparation, 40x40x40 mm specimens of the geopolymer-based mortar mix were cured in molds for 1 day at t = 20 ± 2°C and RH = 55 ± 5%, then removed from the molds and stored for the further 27 days. Compressive strength values have been determined in accordance with EN 196–1.

A coefficient of water resistance of the geopolymer was calculated as a ratio of the values of compressive strength of the specimens at the specified age of 7 days (after preparation, the specimens were kept in molds for 1 day and then in water until the specified age), to those of the specimens hardened to the specified age under normal conditions. It is accepted that a material is classified as a water-resistant material, when the coefficient of water resistance$\geq$0.8.

The bond strength of the geopolymer-based coating to a concrete substrate was determined in accordance with the National Standard of Ukraine DSTU B V.2.6–181 by measuring the perpendicular tensile force required to separate the coating from the concrete substrate using a bonded metal stamp and dynamometer.

Water impermeability (W) of the coated concrete was evaluated in accordance with the National Standard of Ukraine DSTU B V.2.6–181. The thickness of the protective coating was 2 mm. The coated cylinder concrete specimens of Ø150 × 50 mm

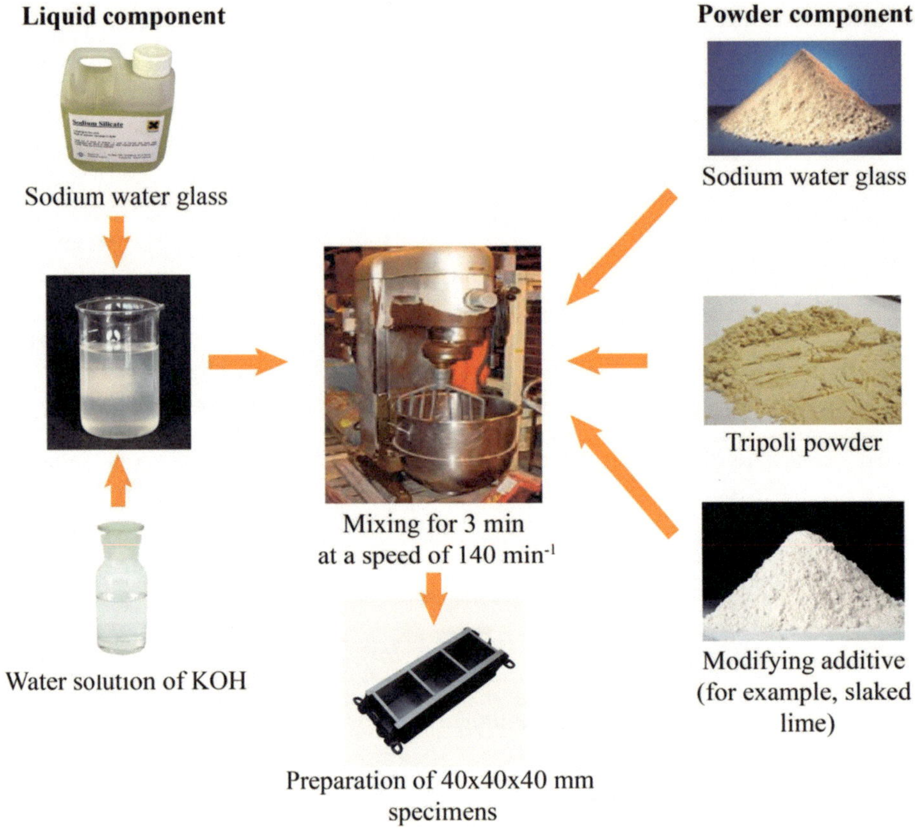

Figure 1.
A scheme of the preparation procedure of geopolymer-based mortars.

that hardened at t = 20 ± 2°C and RH = 95 ± 5% were subjected at the age of 28 days to the action of water under pressure. The pressure was increased in steps, each step being – 0.2 MPa. The storage time at each stage was 6 hr. The grade of concrete in water impermeability is a maximal pressure of water ($MPa \cdot 10^{-1}$) at which no water penetration through the specimen was observed.

The modeling of changes in the protective performances of a geopolymer-based coating used to protect the reinforced concrete, exposed to aggressive environments with a combination of exposure classes XS3 and XC4 as well as with exposure class XS2 (for comparison), was carried out using the author's methodology.

The geopolymer-based coating mix was applied on the side surfaces of the 40 × 40 × 160 mm prism specimens of the fine aggregate concrete (cement to sand ratio 1:3) in thicknesses of ≥5 mm. The coatings were stored for 28 days at t = 20 ± 2°C and RH = 55 ± 5%. A schematic diagram showing the preparation procedure of coated specimens is given in **Figure 2**.

Afterwards, the storage conditions of the coated and uncoated specimens (for comparison) were as follows:

- Periodical wetting (submergence) for 21 hours in seawater and drying for 3 hours (t = 105 ± 3°C) under the influence of atmospheric concentration of CO_2 (K ≈ 0.04%) to model the environment with a combination of exposure classes XS3 and XC4;

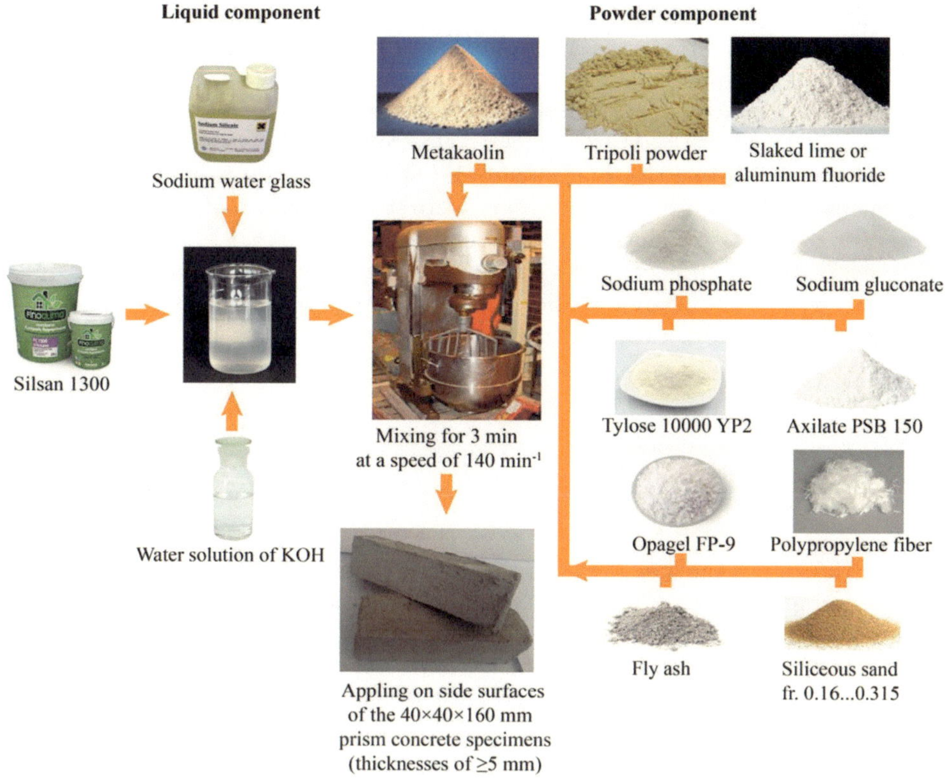

Figure 2.
A scheme of the preparation procedure of coated specimens.

- Permanently submergence in seawater to model the environment with exposure class XS2.

The control terms used to determine the protective performances of the geopolymer-based coating were as follows:

- 50 and 200 cycles of wetting-drying cycles for the environment with a combination of exposure classes XS3 and XC4;
- 3 and 12 months for the environment with exposure class XS2.

Specified control terms were chosen in accordance with the methodology for corrosion testing of steel reinforcement in concrete per National Standard of Ukraine DSTU B V.2.6–171:2008.

The protective performances of the coating were assessed in terms of water absorption, diffusion permeability of Cl^- and SO_4^{2-}-ions, bond strength to a concrete substrate and the state of the steel reinforcement in the concrete.

The water absorption of the coating was determined per National Standard of Ukraine DSTU Б V.2.7–126:2011 by measuring the ratio of the water content absorbed by Karsten tube after 24 h to the area of the material in contact with water. The material is considered to be waterproof if the water absorption is up to 1 kg/m^2 as per the regulatory requirements.

The diffusion permeability of Cl^- and SO_4^{2-}-ions was determined by qualitative reactions due to the spraying of water solutions of silver nitrate $AgNO_3$ (concentration 0.1 N) and lead acetate $Pb(CH_3COO)_2$ (concentration 0.5 N) on the surface of coated and uncoated specimens immediately after their breaking. The depth of penetration of the specified ions was measured using an electronic microscope equipped with microanalyzer REMMA 102–02 (JSC "SELMI", Sumy, Ukraine) to confirm the results of the qualitative reactions. The specimens were cut into slices to determine a depth of penetration of aggressive ions per pointed method.

The evaluation of contact zone state between the coating and concrete was assessed using a Dino-Lite digital microscope (x500 magnification) (AnMo Electronics Corporation, Taiwan).

3. Results

3.1 The efficiency of modifying additives

The efficiency of the modifying additives was evaluated by the changes in compressive strength and water resistance coefficient of the modified geopolymer-based mortars (Mix 2 from **Table 4**) compared to the analog without the additive (Mix 1 from **Table 4**) after 7 days. The consistence (flow) of geopolymer-based mortars mixes was in the range of 18 ± 1 cm. Compressive strength of additive-free hardened mortar (Mix 1) was 35.7 MPa after 7 days and was not water resistant (water resistance coefficient < 0.8).

The application of slaked lime as a *Ca-containing additive* resulted in an increase in compressive strength by 8.1% compared to Mix 1 (**Figure 3**). At the same time, the modification with the additive of $Ca(OH)_2$ caused an increase in the water resistance coefficient from 0.75 up to 0.97 (**Figure 4**). The effects of calcium-containing additives are in compliance with [38] and ensure the formation of the water-resistant stone due to the synthesis of zeolite-like hydroaluminosilicates C-N(K)-A-S-H at normal temperatures, such as gismondite ($CaSi_2Al_2O_8{\cdot}4H_2O$) and thompsonite ($NaCa_2(Al_5Si_5O_{20}){\cdot}6H_2O$).

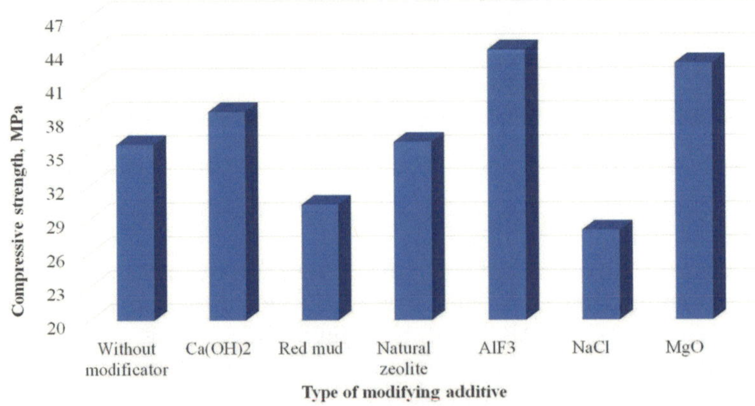

Figure 3.
The compressive strength of geopolymer-based mortar after 7 days versus the type of modifying additive.

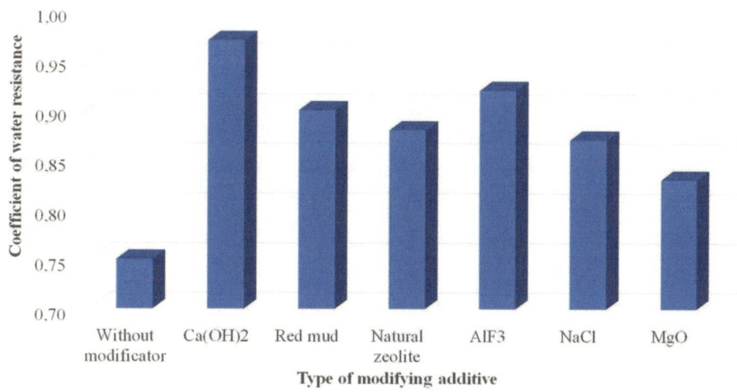

Figure 4.
The coefficient of water resistance of geopolymer-based mortar at 7 days versus the type of modifying additives.

Additive of AlF_3 provides the greatest effect in increasing compressive strength—by 23.5%. The water resistance coefficient of the geopolymer-based mortar also increased up to 0.92. This value of the water resistance coefficient was comparable to that obtained by modifying with $Ca(OH)_2$. The obtained results are consistent with the concept regarding the intensification of the formation of zeolite-like structures C-N(K)-A-S-H with the participation of Al^{3+} ions, accompanied by an increase in strength and water resistance [36]. The formation of cryolite Na_3AlF_6 can also occur with participation of ions F^+ in structure formation processes of geopolymer [44]. The synthesis of cryolite at the early stages of hydration as a film ensures that the setting times of the geopolymer are regulated by microencapsulation of silicate grains.

The possibility of ensuring the water resistance of the mortars was observed when using an additive of *red mud* with composition represented by the oxides Fe_2O_3 + FeO, Al_2O_3, CaO, and SiO_2. A slight decrease in compressive strength of the mortar after 7 days by 15.1% was observed compared to Mix 1. At the same time, the water resistance coefficient of the mortar increased from 0.75 up to 0.90. The increase in water resistance was due to the participation of Ca^{2+}, Al^{3+}, and Si^{4+}-ions, which is evidence of the formation of zeolite-like matrices of the C-N(K)-A-S-H type.

The formation of a waterproof stone was also observed when using the additive of *clinoptilolite*. The water resistance increased from 0.75 up to 0.99 compared to Mix 1. The intensification in the formation of zeolite-like matrices due to the crystal-chemical affinity with the modifying additive caused this effect [55]. There was no change in compressive strength of the geopolymer-based mortar after 7 days.

Magnesium oxide (MgO) as a modifying additive increased the strength of the mortar by 20.2% compared to Mix 1. The water resistance coefficient of the mortar increased up to 0.83 but was not significantly higher than the required value (0.80).

The addition of *sodium chloride (NaCl)* also increased the water resistance coefficient of the mortars (up to 0.87). However, the compressive strength of the mortar decreased by 21.6% after 7 days when using this additive.

According to the obtained results, all the modifying additives cause the formation of the water-resistant stone and can be used as an alternative to those Ca-containing in the geopolymer-based coatings, despite a slight decrease in strength in the case of red mud and NaCl.

3.2 The efficiency of geopolymer-based coatings

The effectiveness of the coatings under wetting-drying cycles (the combination of exposure classes XS3 and XC4) was determined using $Ca(OH)_2$ as a modifying additive (Mix 3 from **Table 5**) and the model of AlF_3 production waste (Mix 4 from **Table 5**). The content of $Ca(OH)_2$ was comparable to the content of the AlF_3 production waste model in terms of AlF_3–0.77% by weight.

The effect of the environment with exposure class XS2 on protective preferences of the geopolymer-based coating was studied for comparison.

The compositions of the geopolymer-based coatings were characterized by technological and performance properties specified in the standards: consistency (flow) –≥ 20 cm, consistency retention –≥ 90 min, water retention –≥ 97%, flexural strength –≥ 3.5, 6.5 i 8.0 MPa after 2, 7 i 28 days, compressive strength –≥ 20.0, 37.0 i 45.0 MPa after 2, 7 i 28 days, the bond strength to a concrete substrate –≥ 1.2 MPa, grade in water impermeability – W10.

3.2.1 Water absorption

The geopolymer-based coating modified with additive of $Ca(OH)_2$ was characterized by an increase in water absorption from 0.0 kg/m^2 to 0.3 kg/m^2 and 0.5 kg/m^2 respectively after 3 months and 12 months of submergence in seawater compared to the coating that had not been subjected to aggressive exposure (reference) (**Figure 5**). An increase in water absorption from 0.0 kg/m^2 up to 0.5 kg/m^2 and 0.8 kg/m^2 was observed after 50 and 200 wetting-drying cycles compared to the reference coating.

The modification of the coating by model of AlF_3 production waste provides water absorption values comparable to those with additive $Ca(OH)_2$. The values of water absorption were 0.5 kg/m^2 and 0.8 kg/m^2 after 3 months and 12 months of full submergence, and 0.8 kg/m^2 and 1.0 kg/m^2 after 50 and 200 wetting-drying cycles.

The obtained values of water absorption (≤ 1 kg/m^2) indicate high waterproofing properties of the geopolymer-based coatings. This is due to the synthesis of water-resistant zeolite-like matrices using modifying additives, which contributes to the denser coating structure.

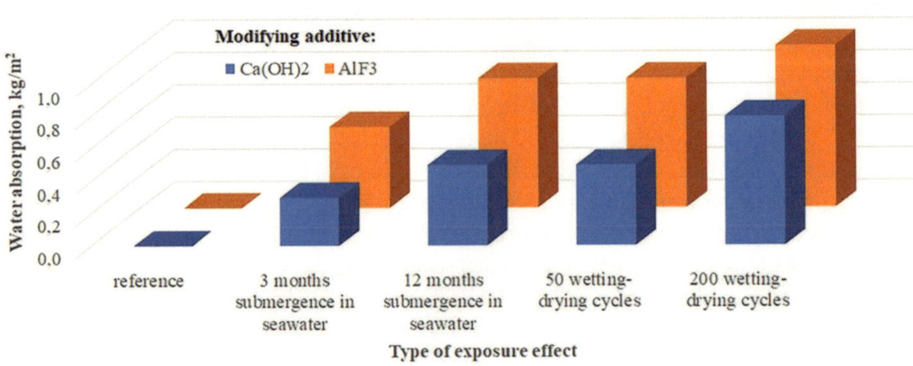

Figure 5.
Water absorption of geopolymer-based coatings versus type of exposure effect.

3.2.2 Bond strength

The bond strength of the geopolymer-based coating to a concrete substrate when using both $Ca(OH)_2$ and the model of AlF_3 production waste, after 3 months and 12 months of submergence in seawater, remained at the level of the coating without aggressive exposure (reference coating)—1.3 MPa and 1.2 MPa (**Figure 6**).

The bond strength of the coating to concrete substrate when using the additive of $Ca(OH)_2$ slightly decreased after 50 and 200 wetting-drying cycles from 1.3 MPa down to 1.2 MPa and 1.1 MPa compared to the reference coating. Almost the same decrease in the bond strength under similar exposure conditions was observed when using the model of AlF_3 production waste—down to 1.1 MPa and 1.0 MPa, respectively.

The observed decrease in the bond strength of geopolymer-based coatings when exposed to aggressive environments is within acceptable limits. The retention of the adhesive properties of the developed coatings is due to the high density of the "coating-concrete" contact zone (**Figure 7**).

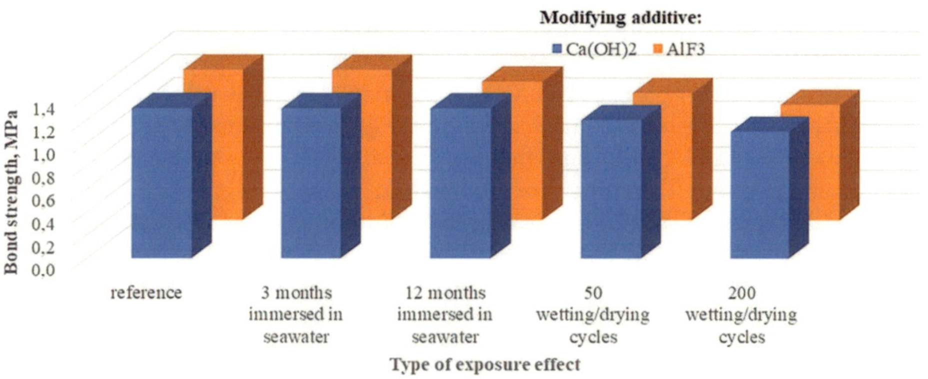

Figure 6.
Bond strength of geopolymer-based coatings to concrete substrate versus type of exposure effect.

Figure 7.
SEM images of "coating-concrete" contact zone when using $Ca(OH)_2$ (a) and model of AlF_3 production waste (b) as modified additives after 200 wetting-drying cycles.

3.2.3 Diffusion permeability of Cl^- and SO_4^{2-}-ions

The diffusion permeability of Cl^- and SO_4^{2-}-ions in uncoated concrete exposed to 50 wetting-drying cycles was 3 mm and 2 mm, respectively, according to the results of the qualitative reactions (Eqs. 1 and 2). The diffusion permeability of Cl^- and SO_4^{2-}-ions in uncoated concrete exposed to 200 wetting-drying cycles was 6 mm (**Figure 8a**) and 3 mm (**Figure 8b**), respectively:

$$NaCl + AgNO_3 \rightarrow AgCl\downarrow + NaNO_3 \quad (1)$$

$$Na_2SO_4 + Pb(CH_3COO)_2 \rightarrow NaCH_3COO + PbSO_4\downarrow \quad (2)$$

The results of the microprobe analysis confirmed the results of the qualitative reactions. The content of Cl^- and SO_4^{2-}-ions (further, referred to as SO_3^{2-}-ions) at a depth of 3 mm of unprotected concrete, exposed to 200 wetting-drying cycles, was 14.40% and 2.35%, respectively (**Figure 9**).

The diffusion permeability of both Cl^- (**Figure 10a**) and SO_4^{2-}-ions (**Figure 10b**) into the $Ca(OH)_2$ modified coating, exposed to 3 months and 12 months of full submergence in seawater (exposure class XS2), was 1 mm. The same values of diffusion permeability of Cl^- (**Figure 11a**) and SO_4^{2-}-ions (**Figure 11b**) into the coating were obtained under the same conditions when the model of AlF_3 production waste was used as a modifying additive. The content of Cl^- and SO_4^{2-}-ions at a depth of 1 mm in the coating, modified with $Ca(OH)_2$ and exposed to 12 months of full submergence in seawater, was 0.50% and 0.24%, respectively, according to the results of the microprobe analysis (**Figure 12a**). The content of Cl^- and SO_4^{2-}-ions at a depth of 1 mm in the coating, modified with the model of AlF_3 production waste and exposed to 12 months of full submergence in seawater, was 0.53% and 0.48%, respectively (**Figure 12b**).

(a) (b)

Figure 8.
SEM of the cleavage surface of half-uncoated specimens, exposed to 200 wetting-drying cycles, after spraying with water solutions of silver nitrate (a) and lead acetate (b).

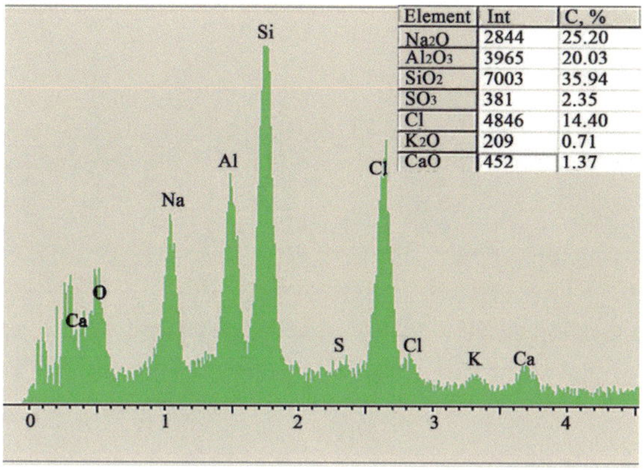

Figure 9.
Microprobe analysis at a depth of 3 mm of unprotected concrete, exposed to 200 wetting-drying cycles.

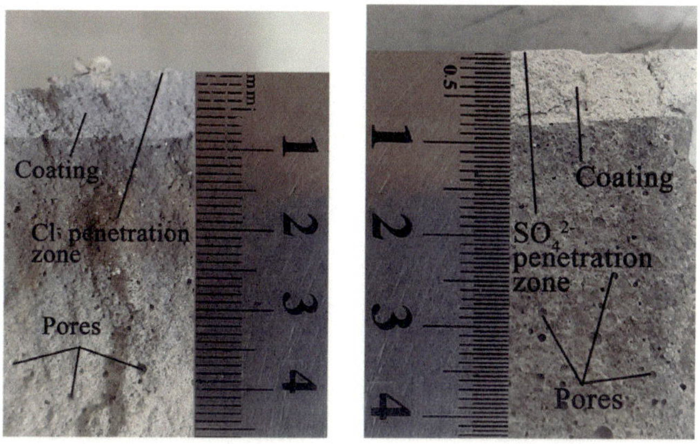

(a) (b)

Figure 10.
SEM of the cleavage surface of half-coated specimens (modifying additive – $Ca(OH)_2$), exposed to 12 months of full submergence in seawater, after spraying with water solutions of silver nitrate (a) and lead acetate (b).

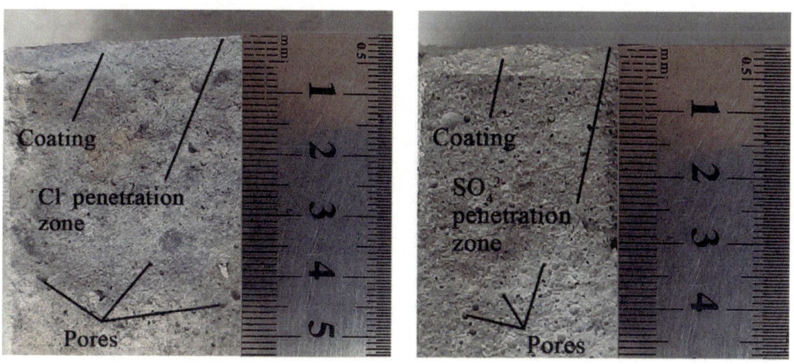

(a) (b)

Figure 11.
SEM of the cleavage surface of half-coated specimens (modifying additive – model of AlF_3 production waste), exposed to 12 months of full submergence in seawater, after spraying with water solutions of silver nitrate (a) and lead acetate (b).

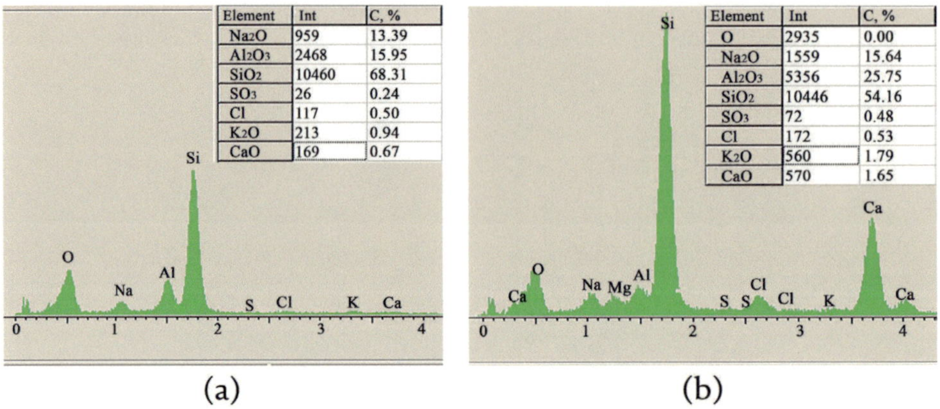

Figure 12.
Microprobe analysis at a depth of 1 mm of the coating, modified with $Ca(OH)_2$ (a) and model of AlF_3 production waste (b) after 12 months of full submergence in seawater.

Higher diffusion permeability was observed when the coating was exposed to aggressive environments with a combination of exposure classes XS3 and XC4. The diffusion permeabilities of Cl^-- and SO_4^{2-}-ions when using the additive of $Ca(OH)_2$ were 1.0 mm and 0.5 mm and were higher in case of the model of AlF_3 production waste – 1.5 mm and 2.0 mm after 50 wetting-drying cycles. After 200 wetting-drying cycles, the diffusion permeabilities of Cl^-- and SO_4^{2-}-ions when using $Ca(OH)_2$ increased up to 1.5 mm (**Figure 13a**) and 0.8 mm (**Figure 13b**) and up to 2.0 mm (**Figure 14a**) and 1.5 mm (**Figure 14b**) when using the model of AlF_3 production waste.

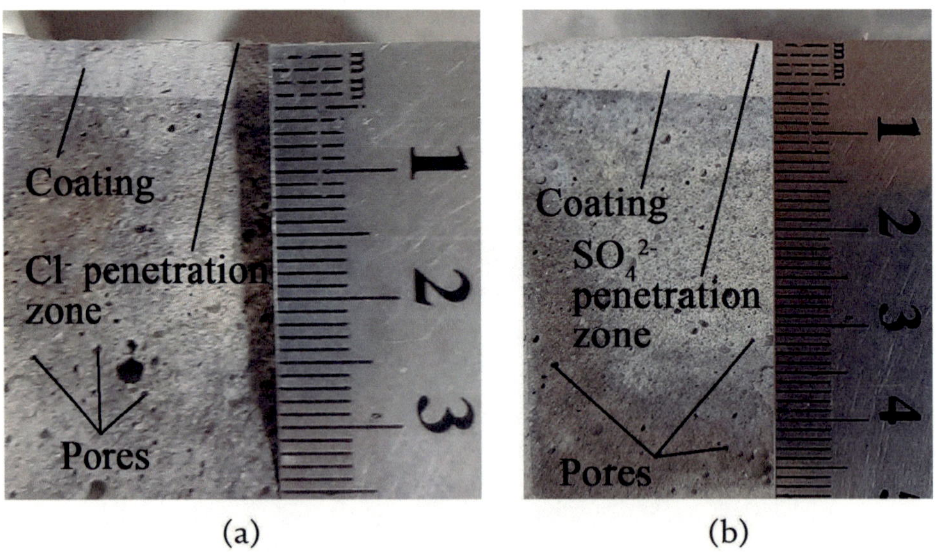

Figure 13.
SEM of the cleavage surface of half-coated specimens (modifying additive – $Ca(OH)_2$), exposed to 200 wetting-drying cycles, after spraying with water solutions of silver nitrate (a) and lead acetate (b).

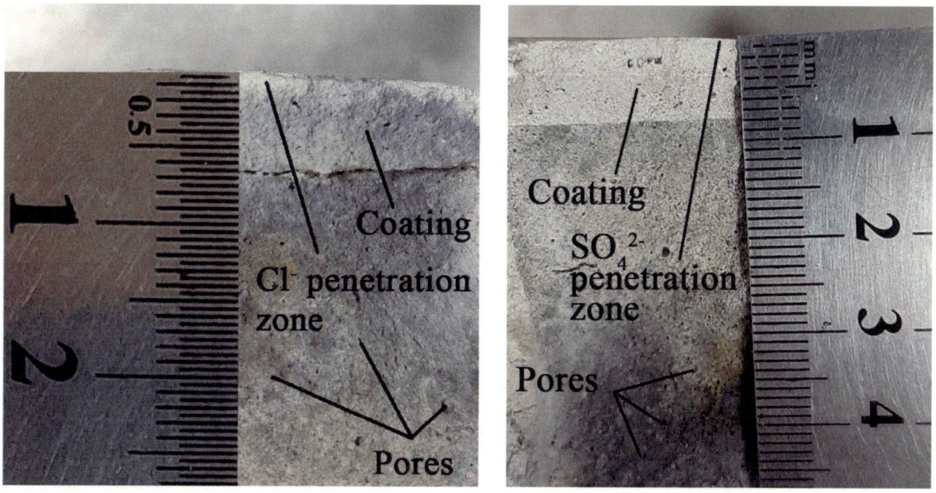

(a) (b)

Figure 14.
SEM of the cleavage surface of half-coated specimens (modifying additive – model of AlF_3 production waste), exposed to 200 wetting-drying cycles, after spraying with water solutions of silver nitrate (a) and lead acetate (b).

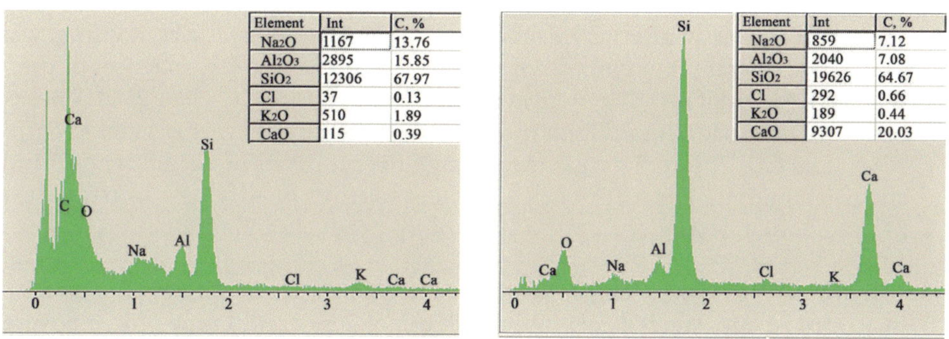

Element	Int	C, %
Na_2O	1167	13.76
Al_2O_3	2895	15.85
SiO_2	12306	67.97
Cl	37	0.13
K_2O	510	1.89
CaO	115	0.39

Element	Int	C, %
Na_2O	859	7.12
Al_2O_3	2040	7.08
SiO_2	19626	64.67
Cl	292	0.66
K_2O	189	0.44
CaO	9307	20.03

(a) (b)

Figure 15.
Microprobe analysis at a depth of 2 mm of the coating, modified with $Ca(OH)_2$ (a), and at a depth of 1.5 mm of the coating, modified with the model of AlF_3 production waste (b), after 200 wetting-drying cycles.

The content of Cl^- and SO_4^{2-}-ions at a depth of 2 mm in the coating, modified with $Ca(OH)_2$ and exposed to 200 wetting-drying cycles, was 0.13% and 0.00%, respectively (**Figure 15a**). At the same time, the content of Cl^- and SO_4^{2-}-ions at a depth of 1.5 mm in the coating, modified with the model of AlF_3 production waste and exposed to 200 wetting-drying cycles, was 0.53% and 0.00%, respectively (**Figure 15b**).

Thus, the dynamics in the development of the permeability were damped when using both $Ca(OH)_2$ and the model of AlF_3 production waste as additives (**Figure 16**).

Thus, the maximum diffusion permeability of Cl^- and SO_4^{2-}-ions into concrete protected by geopolymer-based coating and exploited under seawater conditions was not more than 1...2 mm.

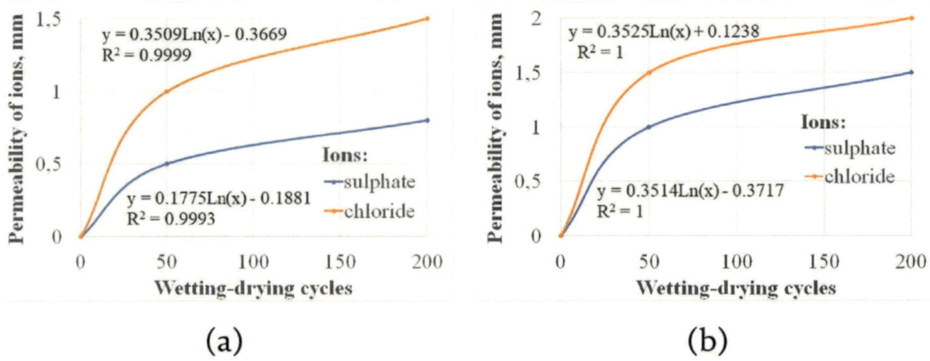

Figure 16.
Permeability of aggressive ions after wetting-drying cycles versus type of modifying additive: $Ca(OH)_2$ (a) and model of AlF_3 production waste (b). Remarks. R^2 – coefficient of determination (statistic).

4. Conclusions

1. The effectiveness of the coating based on the geopolymer $(0{,}2K_2O + 0{,}8Na_2O)\cdot 4{,}5SiO_2\cdot Al_2O_3\cdot nH_2O$ was confirmed for protection of the concrete surface under the cyclic wetting in seawater and drying under the influence of atmospheric CO_2. The addition of salts (AlF_3 and NaCl), metal oxides (MgO and Fe_2O_3), and zeolite (clinoptilolite) as an alternative modifying additive to those Ca-containing, was found to accelerate the synthesis of water-resistant zeolite-like matrices of the $Na(K)_2O$-Al_2O_3-SiO_2-H_2O system at normal temperatures ($t = 20 \pm 2$°C). These structure formation causes limitations in the transport of aggressive ions (Cl^-, SO_4^{2-}, and CO_3^{2-}) into the concrete through the processes of chemisorption and chemical binding. The mentioned substances such as modifying admixtures can be used both independently and as part of by-products and production wastes.

2. The necessity in the coating is due to the diffusion permeability of Cl^- and SO_4^{2-}-ions into the unprotected concrete to a depth of 2 mm and 1 mm – for exposure class XS2, as well as 6 mm and 3 mm – for a combination of exposure classes XS3 and XC4. In the case of protected concrete, the environment with the combination of exposure classes XS3 and XC4 (after 200 cycles of wetting in seawater and their drying) demonstrated a higher degree of aggressiveness compared to the environment with the exposure class XS2 (after 12 months of permanently submergence of concrete specimens in seawater) with 25...60% higher values for water absorption and 50% higher diffusion permeability, and 15...17% lower bond strength to the concrete substrate.

3. The durability of the protective coating was confirmed by the retention of adhesion and water absorption after 200 cycles of wetting in seawater and drying. The high density of the "coating-concrete" contact zone ensures that the adhesion to the concrete substrate is maintained. The water absorption values (≤ 1 kg/m^2) indicate high waterproofing properties of the developed coating. It was predicted that after about 500 cycles, the permeability of Cl^- and SO_4^{2-}-ions in the coating hardly changes, that is, the process is damped.

Author details

Pavlo Kryvenko*, Igor Rudenko and Oleksandr Konstantynovskyi
Scientific Research Institute for Binders and Materials, Kyiv National University of Construction and Architecture, Kyiv, Ukraine

*Address all correspondence to: pavlo.kryvenko@gmail.com

References

[1] Hu J, Zhang S, Chen E, Li W. A review on corrosion detection and protection of existing reinforced concrete (RC) structures. Construction and Building Materials. 2022;**325**:126718. DOI: 10.1016/j.conbuildmat.2022.126718

[2] Qiu Q. A state-of-the-art review on the carbonation process in cementitious materials: Fundamentals and characterization techniques. Construction and Building Materials. 2020;**247**:118503. DOI: 10.1016/j.conbuildmat.2020.118503

[3] Krivenko P, Rudenko I, Konstantynovskyi O, Boiko O, Vaičiukynienė D. Effect of sodium phosphate and sodium nitrate on microstructure of alkali-activated slag cement pastes and properties of reinforced concrete under cyclic drying-wetting in sea water. AIP Conference Proceedings. 2023;**2840**:020006-1-020006-11. DOI: 10.1063/5.0168007

[4] Qu F, Li W, Dong W, Tam VWY, Yu T. Durability deterioration of concrete under marine environment from material to structure: A critical review. Journal of Building Engineering. 2021;**35**:102074. DOI: 10.1016/j.jobe.2020.102074

[5] Roziere E, Loukili A, El HR, Grondin F. Durability of concrete exposed to leaching and external sulphate attacks. Cement and Concrete Research. 2009;**39**(12):1188-1198. DOI: 10.1016/j.cemconres.2009.07.021

[6] Jones C, Ramanathan S, Suraneni P, Hale WM. Calcium oxychloride: A critical review of the literature surrounding the formation, deterioration, testing procedures, and recommended mitigation techniques. Cement and Concrete Composites. 2020;**113**:103663. DOI: 10.1016/j.cemconcomp.2020.103663

[7] Qiao C, Suraneni P, Weiss J. Flexural strength reduction of cement pastes exposed to $CaCl_2$ solutions. Cement and Concrete Composites. 2018;**86**:297-305. DOI: 10.1016/j.cemconcomp.2017.11.021

[8] Wu J, Zhu W, Wei J, Tang M, Zhang S, Zhang X. A comparable study on the deterioration of concrete under sodium sulfate and magnesium sulfate attack. IOP Conference Series: Earth and Environmental Science. 2021;**787**:012037. DOI: 10.1088/1755-1315/787/1/012037

[9] Khan MU, Ahmad S, Al-Gahtani HJ. Chloride-induced corrosion of steel in concrete: An overview on chloride diffusion and prediction of corrosion initiation time. International Journal of Corrosion. 2017;**2017**:1-9. DOI: 10.1155/2017/5819202

[10] Mundra S, Bernal SA, Criado M, et al. Steel corrosion in reinforced alkali-activated materials. RILEM Technical Letter. 2017;**2**:33-39. DOI: 10.21809/rilemtechlett.2017.39

[11] Andrade C. Some historical notes on the research in corrosion of reinforcement. Hormigón y Acero. 2018;**69**(S1):21-28. DOI: 10.1016/j.hya.2018.12.002

[12] Wang Z, Zhang L, Tang X, Cui Z, Xue J, Lu M. Investigation of the deterioration of passive films in H_2S-containing solutions international journal of minerals. Metallurgy and Materials. 2017;**24**(8):943-953. DOI: 10.1007/s12613-017-1482-6

[13] EN 206:2013 Concrete - Specification, performance, production and conformity

[14] Safehian M, Ramezanianpour AA. Assessment of service life models for determination of chloride penetration into silica fume concrete in the severe marine environmental condition. Construction and Building Materials. 2013;**48**:287-294. DOI: 10.1016/j.conbuildmat.2013.07.006

[15] Xu L, Zhang Y, Zhang S, Fan S, Chang H. Effect of carbonation on chloride maximum phenomena of concrete subjected to cyclic wetting–drying conditions: A numerical and experimental study. Materials. 2022;**15**:2874. DOI: 10.3390/ma15082874

[16] Wu L, Li W, Yu X. Time-dependent chloride penetration in concrete in marine environments. Construction and Building Materials. 2017;**152**:406-413. DOI: 10.1016/j.conbuildmat.2017.07.016

[17] Qi B, Gao J, Chen F, Shen D. Chloride penetration into recycled aggregate concrete subjected to wetting–drying cycles and flexural loading. Construction and Building Materials. 2018;**174**:130-137. DOI: 10.1016/j.conbuildmat.2018.04.122

[18] Zhongya Z, Xiaoguang J, Wei L. Long-term behaviors of concrete under low-concentration sulfate attack subjected to natural variation of environmental climate conditions. Cement and Concrete Research. 2019;**116**:217e230. DOI: 10.1016/j.cemconres.2018.11.017

[19] Jiang L, Niu D. Study of deterioration of concrete exposed to different types of sulfate solutions under drying-wetting cycles. Construction and Building Materials. 2016;**44**(117):88-98. DOI: 10.1016/j.conbuildmat.2016.04.094

[20] Chang H, Mu S, Feng P. Influence of carbonation on “maximum phenomenon” in surface layer of specimens subjected to cyclic drying-wetting condition. Cement and Concrete Research. 2018;**103**:95-109. DOI: 10.1016/j.cemconres.2017.10.005

[21] Malheiro R, Camões A, Meira G, Amorim MT, Castro-Gomes J. Effect of coupled deterioration by chloride and carbonation on chloride ions transport in concrete. RILEM Technical Letter. 2020;**5**:56-62. DOI: 10.21809/rilemtechlett.2020.126

[22] Ye H, Jin X, Fu C, Jin N, Xu Y, Huang T. Chloride penetration in concrete exposed to cyclic drying-wetting and carbonation. Construction and Building Materials. 2016;**112**:457e463. DOI: 10.1016/j.conbuildmat.2016.02.194

[23] Chang HL, Feng P, Lyu K, Liu J. A novel method for assessing C-S-H chloride adsorption in cement pastes. Construction and Building Materials. 2019;**225**:324-331. DOI: 10.1016/j.conbuildmat.2019.07.212

[24] Geng J, Easterbrook D, Liu QF, Li LY. Effect of carbonation on release of bound chlorides in chloride contaminated concrete. Magazine of Concrete Research. 2016;**68**:353-363. DOI: 10.1680/jmacr.15.00234

[25] Sanytsky M, Kropyvnytska T, Ivashchyshyn H. Sustainable modified pozzolanic supplementary cementitious materials based on natural zeolite, fly ash and silica fume. IOP Conference Series: Earth and Environmental Science. 2023;**1254**:012004. DOI: 10.1088/1755-1315/1254/1/012004

[26] Kropyvnytska T, Sanytsky M, Heviuk I, Kripka L. Study of the properties of low-carbon Portland-composite cements CEM II/C-M. Lecture Notes in Civil

Engineering. 2023;**290**:230-237. DOI: 10.1007/978-3-031-14141-6_22

[27] Goyal A, Pouya HS, Ganjian E, Claisse P. A review of corrosion and protection of steel in concrete. Arabian Journal for Science and Engineering. 2018;**43**:5035-5055. DOI: 10.1007/s13369-018-3303-2

[28] Kryvenko P, Sikora P, Rudenko I, Konstantynovskyi O. Advances in using seawater in slag-containing cement systems. Journal of Building Engineering. 2024;**96**:110386. DOI: 10.1016/j.jobe.2024.110386

[29] Hosseinzadeh N, Ghiasian M, Andiroglu E, Lamere J, Rhode-Barbarigos L, Sobczak J, et al. Concrete seawalls: A review of load considerations, ecological performance, durability, and recent innovations. Ecological Engineering. 2022;**178**:106573. DOI: 10.1016/j.ecoleng.2022.106573

[30] Pan X, Shi Z, Shi C, Ling T-C, Li N. A review on concrete surface treatment. Part I: Types and mechanisms. Construction and Building Materials. 2017;**132**:578-590. DOI: 10.1016/j.conbuildmat.2016.12.025

[31] Wang H, Feng P, Lv Y, Geng Z, Liu Q, Liu X. A comparative study on UV degradation of organic coatings for concrete: Structure, adhesion, and protection performance. Progress in Organic Coatings. 2020;**149**:105892. DOI: 10.1016/j.porgcoat.2020.105892

[32] Sánchez M, Faria P, Ferrara L, Horszczaruk E, Jonkers HM, Kwiecien A, et al. External treatments for the preventive repair of existing constructions: A review. Construction and Building Materials. 2018;**193**:435-452. DOI: 10.1016/j.conbuildmat.2018.10.173

[33] Delucchi M, Barbucci A, Cerisola G. Study of the physico-chemical properties of organic coatings for concrete degradation control. Construction and Building Materials. 1997;**11**(7):365-371. DOI: 10.1016/S0950-0618(97)00060-3

[34] Balaguru P, Nazier M, Arafa M. Field Implementation of Geopolymer Coatings. Project Report of Center for Advanced Infrastructure and Transportation (CAIT). Civil and Environmental Engineering Piscataway. NJ: Rutgers State University; 2008

[35] Krivenko P, Guzii S, Bodnarova L, Valek J, Hela R, Zach J. Effect of thickness of the intumescent alkali aluminosilicate coating on temperature distribution in reinforced concrete. Journal of Building Engineering. 2016;**8**:14-19. DOI: 10.1016/j.jobe.2016.09.003

[36] Kryvenko P, Rudenko I, Sikora P, Sanytsky M, Konstantynovskyi O, Kropyvnytska T. Alkali-activated cements as sustainable materials for repairing building construction: A review. Journal of Building Engineering. 2024;**90**:109399. DOI: 10.1016/j.jobe.2024.109399

[37] Krivenko P, Kyrychok V. In: Vizureanu P, Krivenko P, editors. Advances in Geopolymer-Zeolite Composites – Synthesis and Characterization: Monograph. London: IntechOpen; 2021. DOI: 10.5772/intechopen.93360

[38] Krivenko P, Rudenko I, Konstantynovskyi O, Vaičiukynienė D. Feasibility of incorporating SO_4^{2-}-ions in zeolite-like matrices based on alkaline aluminosilicate binders. Construction and Building Materials. *2023*;**391**:131878. DOI: 10.1016/j.conbuildmat.2023.131878

[39] Kryvenko P, Rudenko I, Konstantynovskyi O, Vaičiukynienė D.

Concrete protection from sulfate aggression by coating based on zeolite-like binders. Advances in Transdisciplinary Engineering. 2023;**43**:607-614. DOI: 10.3233/ATDE230772

[40] Franus W, Wdowin M, Franus M. Synthesis and characterization of zeolites prepared from industrial fly ash. Environmental Monitoring and Assessment. 2014;**2014**(186):5721-5729. DOI: 10.1007/s10661-014-3815-5

[41] Klunk M, Dasgupta S, Das M, Cunha MG, Wander PR. Synthesis of sodalite zeolite and adsorption study of crystal violet dye. ECS Journal of Solid State Science and Technology. 2019;**8**(114):144-150. DOI: 10.1149/2.0131910jss

[42] Saifuddin M, Bae J, Kim KS. Role of Fe, Na and Al in Fe-zeolite-a for adsorption and desorption of phosphate from aqueous solution. Water Research. 2019;**158**:246-256. DOI: 10.1016/j.watres.2019.03.045

[43] Krivenko P, Kovalchuk O, Pasko A. Utilization of industrial waste water treatment residues in alkali activated cement and concretes. Key Engineering Materials. 2018;**761**:35-38. DOI: 10.4028/www.scientific.net/kem.761.35

[44] Vaičiukyniene D, Vaitkevičius V, Kantautas A, Sasnauskas V. Effect of AlF_3 production waste on the properties of hardened cement paste. Medziagotyra. 2012;**18**(2):187-191. DOI: 10.5755/j01.ms.18.2.1925

[45] Kovalchuk O, Kochetov G, Samchenko D, Kolodko A. Development of a technology for utilizing the electroplating wastes by applying a ferritization method to the alkaline activated materials. Eastern-European Journal of Enterprise Technologies. 2019;**2**(10 (98)):27-34. DOI: 10.15587/1729-4061.2019.160959

[46] Kochetov G, Prikhna T, Samchenko D, Kovalchuk O. Development of feritization processing of galvanic waste involving the energy saving electromagnetic pulse activation of the process. Eastern-European Journal of Enterprise Technologies. 2019;**6**(10 (102)):6-14. DOI: 10.15587/1729-4061.2019.184179

[47] Vaičiukynienė D, Kantautas A, Nizevičienė D, Vaičiukynas V, Alaburdaitė R. Utilization of polypropylene waste and zeolitic by-product in composite materials. Materials Letters. 2024;**367**:136563. DOI: 10.1016/j.matlet.2024.136563

[48] Venkatesh C, Chand MSR, Nerella R. A state of the art on red mud as a substitutional cementitious material. Annales de Chimie: Science des Materiaux. 2019;**43**(2):99-106. DOI: 10.18280/acsm.430206

[49] Kong L, Han M, Yang X. Evaluation on relationship between accelerated carbonation and deterioration of concrete subjected to a high-concentrated sewage environment. Construction and Building Materials. 2020;**327**:117650. DOI: 10.1016/j.conbuildmat.2019.117650

[50] Kaiyuan M, Tao G, Youzhi Z, Liwei Z, Feng Z, Peng G, et al. Effectiveness and microstructure change of alkali-activated materials during accelerated carbonation curing. Construction and Building Materials. 2021;**274**:122063. DOI: 10.1016/j.conbuildmat.2020.122063

[51] Li Z, Li S. Effects of wetting and drying on alkalinity and strength of fly ash/slag-activated materials. Construction and Building Materials.

2020;**254**:119069. DOI: 10.1016/j.conbuildmat.2020.11906

[52] Beltrame N, Dias R, Witzke F, Medeiros-Junior R. Effect of carbonation curing on the physical, mechanical, and microstructural properties of metakaolin-based geopolymer concrete. Construction and Building Materials. 2023;**406**:133403. DOI: 10.1016/j.conbuildmat.2023.133403

[53] Bernal SA, Provis JL, de Gutiňrrez M, van Deventer JSJ. Accelerated carbonation testing of alkali-activated slag/metakaolin blended concretes: Effect of exposure conditions. Materials and Structures. 2015;**48**(3):653-669. DOI: 10.1617/s11527-014-0289-4

[54] Dickson AG, Sabine CL, Christian JR. Guide to best practices for ocean CO_2 measurement. In: PICES Special Publication 3. Sidney, British Columbia: North Pacific Marine Science Organization; 2007. 191 p

[55] Pushkareva KK, Gonchar OA, Kaverin KO. The role of the crystallo-chemical factor in the evaluation and improvement of the nanomodification efficiency of mortar and concrete. IOP Conference Series: Materials Science and Engineering. 2019;**708**:012102. DOI: 10.1088/1757-899X/708/1/012102